Developing Research Skills

Developing Research Skills
A Laboratory Manual

Third Edition

by

Helen J. Crawford
Virginia Polytechnic Institute and State University

Larry B. Christensen
University of South Alabama

Allyn and Bacon
Boston • London • Toronto • Sydney • Tokyo • Singapore

Copyright © 1995, 1991, and 1985 by Allyn & Bacon
A Simon & Schuster Company
Needham Heights, Massachusetts 02194

ISBN 0-205-15569-3

Printed in the United States of America

10 9 8 7 6 5 4 3 2 1   98 97 96 95 94

CONTENTS

PREFACE

Traditional research laboratory manuals in the Behavioral Sciences emphasize the experimental approach to hypothesis testing. In *Developing Research Methods: A Laboratory Manual* we provide not only the experimental approach but also other methodological approaches: naturalistic observation, archival research, survey research, and correlational research. Thus, like many research textbooks available today we emphasize a multimethod approach to hypothesis testing.

The laboratory exercises within each of the topics provide students with training in how to do both descriptive and experimental research in various settings. The exercises have been designed to train students systematically in how to use and evaluate the scientific method in psychological research. They will learn how to identify a problem and conduct a literature search, to develop hypotheses and translate them into appropriate research designs, to define variables conceptually and operationally, to choose appropriate populations for study in nonbiased manners, to develop control and reduce extraneous variables, to design research, to gather and analyze data, to interpret results, and to communicate findings to others.

This manual serves both future researchers and consumers of research. It is a training ground for students who will go on to more advanced research methodology and psychology courses. Also it trains individuals to be discerning research consumers so that when they read research they will do it with a critical eye, evaluating its appropriateness and limitations.

Rather than emphasize specific content areas, we have chosen exercises that maximize student interest. Some of the exercises permit the students to learn more about themselves (a desire of many students who take psychology courses). The topics include exercises on learning styles, dreams, ESP, alcohol usage among college students, eating disorders among college students, and illusions, to name but a few.

The manual is organized to provide maximum flexibility to the instructor. Alternate laboratory exercises are included in each of the fourteen topics covered in the laboratory manual. Instructors are not expected to use all of the exercises; rather, they can choose those exercises they think most appropriate for their class. For some of the topics in which actual research studies are conducted, it is suggested that different exercises be assigned to group of students within the class. For example, for Topic 6 which addresses survey research, some students could be assigned to carry out a survey on alcohol usage on campus, while others could be assigned to do a survey on the prevalence of bulimia or on attitudes towards ESP.

The presentation order of the topics may be changed. We recommend that Topics 1 (Scientific Observation), 2 (Problem Identification and Hypothesis Testing), and 3 (Operational Definitions) come first in any course. The subsequent topics can be reorganized to fit the needs of the course. Presently the subsequent laboratory exercises are organized from the least controlled (naturalistic observations) to the most controlled (experimental) research approaches. Another logical presentation order would be to present first the topics addressing experimental research (Topics 7, 8, 9, 10, 11 and 12) and secondly those addressing descriptive research (Topics 4, 5, 6, and 7). Topic 14 addresses ethical issues and the development of consent forms and debriefing statements; this can be introduced at the appropriate time in the course. Finally, Topic 15 provides training in the evaluation of research articles, and can be presented as a single exercise or divided into separate exercises to be turned in periodically throughout the class term. Some topics may need to be eliminated due to time constraints. Certain exercises could also be used as supplementary material in the experimental course which accompanies the laboratory.

While many exercises direct the students as to exactly what to do and hand in to the instructor, others indicate that the instructor will tell the student what to write up. It is our experience that some instructors wish the study to be written up completely, while other instructors ask students to write up only one part of the study (for example, introduction, method, results, or discussion sections). It is left up to the instructor to decide how much of the study should be written up.

We recommend strongly that the students either purchase or have available for their use the latest edition of the Publication Manual of the American Psychological Association. In addition, it is expected that the instructor will provide guidance as to which statistics are appropriate for certain exercises. This decision depends upon the statistics background of the students. As students increasingly own their own computers or have access to them on their campuses, a number of instructors have begun to require students to use statistical packages. While you may have your own package already, if not, we recommend any of the following:

1. Abbott, D. W. (1989). *Instructional Support Statistics*. Available from David W. Abbott, Ph.D., Department of Psychology, University of Central Florida, Orlando, FL 32816. ($10 with manual; $49.95 for site license)

2. *MYSTAT: A personal version of SYSTAT* (1988). Evanston, IL: Systat, Inc. Available from SYSTAT, Inc., 1800 Sherman Avenue, Evanston, IL 60201

3. Norusis, M. (1988). *SPSS/PC* Studentware.* Chicago, IL: SPSS. Can be ordered through your bookstore.

We encourage you and your students to give us feedback. If you have any constructive criticisms or suggestions for improvements, please send them to Helen Crawford, Virginia Polytechnic Institute and State University, Blacksburg, VA 24061.

Word of Thanks

We owe a debt of appreciation to the many people who assisted us in revising our laboratory manual. Our own students in experimental courses and the graduate teaching assistants who often oversaw the actual laboratory exercises provided excellent suggestions for improvement. Special appreciation is given to those individuals who provided formal reviews of the earlier edition.

We particularly want to thank Deborah Akers and Helen Salmon for typing an excellent camera-ready manuscript. Their skills in word processing and organization of materials made this edition much more professional looking.

Finally, the superb professionals at Allyn and Bacon were extremely helpful at all stages of the revision.

TOPIC 1: SCIENTIFIC OBSERVATION

Psychologists use the scientific method to acquire knowledge about human and animal behavior. It is a logical process or method of acquiring information that involves five major steps: identifying the problem and forming a hypothesis, designing the research, conducting the research, testing the hypotheses, and communicating the results. A researcher often does not move systematically through these steps, but rather moves back and forth (e.g., Christensen, 1994). When one uses the scientific method rather than other approaches of acquiring knowledge (tenacity, intuition, authority, rationalism and empiricism), the emphasis is upon making objective observations without bias or opinion.

Psychologists use a wide variety of techniques to carry out their scientific observations. They observe behavior either with or without the knowledge of those being observed. They may use archival material, informal observations, standardized observations, self-reports requiring open-ended responses or responses to rating scales, personality tests, cognitive and perceptual tests, neuropsychological tests, experimental tasks, physiological measures, and so on. They may carry out their research in an experimental laboratory, in a clinical setting, in the field, or they may use previously collected material as their data. These many techniques can be categorized under several research design approaches:

1. Descriptive Research Approach. This approach describes accurately the phenomena being studied, but does not attempt to determine the cause-and-effect relationships between variables. Subsumed under this approach are a number of types of research: case studies, naturalistic observation, secondary record studies, field studies, surveys, correlational studies, and ex post facto studies.

2. Experimental Research Approach. This approach attempts to determine the cause-and-effect relationship between variables by using the psychological experiment, which is an "objective observation of phenomena which are made to occur in a strictly controlled situation in which one or more factors are varied and the others are kept constant" (Zimney, 1961, p. 18).

Two major research settings for psychological research are the laboratory and the field, although psychologists also use other settings (e.g., library for archival studies). The laboratory permits the investigator to develop more precise control over the environment, events, tasks, and stimuli to which the research subjects are exposed. As the control of manipulated variables and extraneous variables increases, the researcher can be more confident about any cause-and-effect conclusions that are drawn from the results. Field research, often in a natural setting, provides the advantage of studying people as they naturally behave but often has the disadvantage of less control.

The laboratory exercises within each of the topics provided in this laboratory manual provide you with training in how to do both descriptive and experimental research in various settings. The exercises have been designed to train you systematically in how to use and evaluate the scientific method in psychological research. You will learn how to identify a problem and conduct a literature search, to develop hypotheses and translate them into appropriate research designs, to define your variables conceptually and operationally, to choose appropriate populations for study in nonbiased manners, to develop control and reduce extraneous variables, to design your procedures, to gather and analyze your data, to interpret your findings, and to communicate your findings to others. A good researcher is always improving and refining his or her research approaches. Research should be exciting and rewarding. Certainly there is always a lot of mundane and tedious daily work, as is the case in most careers, but the excitement of acquiring new knowledge and communicating it to others is worth it to the committed researcher.

While you are presently a novice, you will probably be surprised by how much you learn by the end of the course. You may not become a researcher in psychology, but you will become a much more sophisticated consumer of scientific research. When you read newspapers and magazines you will be able to analyze research reports critically and to evaluate the appropriateness and limitations of the research. If you are in a community organization, you will have the basic knowledge of how to design and carry out surveys and to communicate their results to others.

With the guidance of our own students in past research methods classes we, the authors of this laboratory manual, have chosen and refined what we think to be interesting and relevant exercises. So often we have had psychology students complain that they do not learn enough about themselves in the process of studying psychology (and, yet, that is one of the major reasons many students take psychology courses!). For that reason, we have included several exercises that are aimed at learning more about yourself. The first laboratory exercise below is just such an exercise.

Psychology has conducted thousands of experiments on learning and retention of material and developed major theories about learning and memory. While many of us think we are efficiently reading and retaining information, we may not be as efficient as we could be. The first exercise is aimed at helping you understand your own learning style, its strengths and weaknesses, while at the same time demonstrating that science requires objective observations. A comparison of learning strategies with grade point average will be made. While the exercise is correlational in nature, we hope it will encourage you to think of how you might change your learning strategies to improve your ability to gain knowledge (and improve your grades!).

Laboratory Exercises 2 and 3 provide additional examples illustrating the necessity of objective observation by involving you as an active participant in the recall of your own dreams.

References:

Christensen, L. B. (1994). *Experimental psychology* (6th ed.). Boston: Allyn and Bacon, Inc.
Zimney, G. H. (1961). *Method in experimental psychology*. New York: Ronald Press.

LABORATORY EXERCISE 1: RELATIONSHIP BETWEEN LEARNING STYLES AND GRADE POINT AVERAGE

The purpose of this laboratory exercise is to determine the relationship between various learning styles and cumulated grade point average.

Before proceeding with the reading of this laboratory exercise, turn to the end of this exercise. Fill out the Inventory of Learning Processes (Schmeck, 1983) as honestly as you can. When you have completed the inventory, turn to the following page in this exercise where directions are given as to how it is to be scored. Score your inventory and anonymously turn your scores into your instructor.

We do not all study or comprehend information in the same manner. Different students use different strategies when reading or listening to information. Some individuals tend to think deeply and elaboratively while others think shallowly and reiteratively while reading, studying, or listening to a lecture. Human learning and memory research suggests that individuals who process information deeply and elaboratively recall more information than those who do not. Craik and Lockhart (1972) first proposed that there were levels of processing, falling on a continuum from shallow (information is repeated) to deep (meanings and associations are evaluated).

Not only do individuals do better on laboratory tasks if they process information deeply and elaboratively, but also recent research indicates that such a relationship is found in applied settings, such as classrooms. Several inventories have been developed which assess various learning styles. The one we are interested in, the Inventory of Learning Processes, was developed by Ronald Schmeck and his colleagues at Southern Illinois University. This inventory is made up of 62 true-false statements concerned with learning activities within the school environment. The inventory is made up of four scales, combining individual statements. These four scales, reflecting different learning styles, are the following:

(1) <u>Deep Processing</u>. Consisting of 18 items, this scale assesses "the extent to which students critically evaluate, conceptually organize, and compare and contrast the information they study" (Schmeck, 1983, p. 245).

(2) <u>Elaborative Processing</u>. Consisting of 14 items, this scale assesses "the extent to which students translate new information into their own terminology, generate concrete examples from their own experience, apply new information to their own lives, and use visual imagery to encode new ideas" (Schmeck, 1983, p. 248).

(3) <u>Fact Retention</u>. Consisting of 7 items, this scale assesses how carefully individuals "process (and thus store) details and specific pieces of new information regardless of what other information-processing strategies they might employ" (Schmeck, 1983, p. 248).

(4) <u>Methodical Study</u>. Consisting of 23 items, this scale assesses the degree to which individuals "claim to study more often and more carefully than other students, and the methods that they claim to employ (that) are the systematic techniques recommended in all of the old 'how to study' manuals (e.g., 'type your notes, outline the text, study everyday in the same location, make up practice tests, etc.')" (Schmeck, 1983, p. 249).

The intercorrelations between the four scales indicate that Deep Processing and Elaborative Processing are the most closely related. Craik and Tulving (1975) argue that these are the most important ways to improve memory. They do differ somewhat in that they predict different, although related, performances. Schmeck (1983) reported that various studies show that Deep Processing is related to critical thinking ability, reading comprehension, verbal ability, attention to the semantic attributes of words, and the Wechsler Adult Intelligence Test's digit-span subtest. The Elaborative Processing scale is related to writing performance, use of mental imagery, subjective organization of recalled word lists, and the tendency to organize work lists around rhymes. Schmeck (1983) reported that "the person who scores high on Fact Retention is prone to follow instructions carefully, to be bound by the course syllabus, and to process details, while the person who scores high on Elaborative Processing is able to elaborate and personalize information verbally as well as through imagery" (p. 252). Finally, Schmeck and his colleagues have found that the person who scores high on Methodical Study is achievement striving "of the <u>conforming</u> sort (which) suggests that these students are eager to please and are bound by the course syllabus" (pp. 251-252).

College students who have high grade point averages (GPAs) score significantly higher on the Deep Processing subscale than those with average or low GPAs (Bartling, 1988; Gadzella, Ginther, & Williamson, 1986; Miller, Alway, & McKinley, 1987: Schmeck & Grove, 1979). Students with high GPAs and high American College Test (ACT) assessment scores tend to score high on deep processing, elaborative processing, and fact retention (Schmeck & Grove, 1979). While GPA was not correlated with Methodical Study, low ACT scorers scored significantly higher on Methodical Study than did high ACT scorers. Such relationships between these learning styles and academic performance is not attributable to intelligence differences (Schmeck, 1983).

For your laboratory exercise, compute the correlation between the four learning style scales and cumulated grade point average for the students in your class. Below are some means for each of the scales, taken from Schmeck (1983). Compare your own scores and the class' means with them.

	Deep Processing	Elaborative Processing	Fact Retention	Methodical Study
Low GPA students	10.40	9.80	4.26	10.15
High GPA students	12.58	10.53	5.11	10.57

Your instructor may require you to read some of the references provided below. Schmeck, Geisler-Brenstein, and Cercy (1991) have revised the Inventory of Learning Process. You may wish to explore how it has been revised and expanded.

As a class discussion, you may wish to discuss methods that might improve your learning habits and, hopefully, your academic performance.

Your instructor will direct you as to how to write up this exercise.

Learning Style References:

Bartling, C. A. (1988). Longitudinal changes in the study habits of successful college students. *Educational and Psychological Measurement, 48*, 527-535.

Beyler, J., & Schmeck, R. R. (1992). Assessment of individual differences in preferences for holistic-analytics strategies: Evaluation of some commonly available instruments. *Educational and Psychological Measurement, 52*, 709-719.

Craik, F. I. M., & Lockhart, R. S. (1972). Levels of processing: A framework for memory research. *Journal of Verbal Learning and Verbal Behavior, 11*, 671-684.

Gadzelle, B. M., Ginther, D. W., Williamson, J. D. (1986). Differences in learning processes and academic achievement. *Perpetual and Motor Skills, 62*, 151-156.

Hermann, D. J. (1982). Know they memory: The use of questionnaires to assess and study memory. *Psychological Bulletin, 92*, 434-452.

McCarthy, P. R., Shaw, T., & Schmeck, R. R. (1986). Behavioral analysis of client learning during counseling. *Journal of Counseling Psychology, 33*, 249-254.

Miller, C. D., Alway, M., & McKinley, D. L. (1987). Effects of learning styles and strategies on academic success. *Journal of College Student Personnel, 28*, 399-404.

Schmeck, R. R. (1983). Learning styles of college students. In R. Dillon, & R. Schmeck (Eds.), *Individual differences in cognition*. New York: Academic Press.

Schmeck, R. R., & Geisler-Brenstein, E. (1989). Individual differences that affect the way students approach learning. *Learning and Individual Differences, 1*, 85-124.

Schmeck, R. R., Geisler-Brenstein, E., & Cercy, S. P. (1991). Self-concept and learning: The revised Inventory of Learning Processes. *Educational Psychology, 11*, 343-362.

Schmeck, R. R., & Grove, E. (1979). Academic achievement and individual differences in learning processes. *Applied Psychological Measurement, 3*, 43-49.

Schmeck, R. R., & Meier, S. T. (1984). Self-reference as a learning strategy and a learning style. *Human Learning, 3*, 9-17.

Schmeck, R. R., & Phillips, J. (1982). Levels of processing as dimension of difference between individuals. *Human Learning, 1*, 95-103.

Schmeck, R. R., Ribich, F. D., & Ramanaiah, N. (1977). Development of a self-report inventory for assessing individual differences in learning processes. *Applied Psychological Measurement, 1*, 413-431.

Inventory of Learning Processes

This questionnaire asks you to describe the way you study and learn. There are many different ways to study and learn, any of which may be effective for a particular individual. Since this is the case, there are no "right" or "wrong" answers to these questions. We are simply trying to find out the ways in which people learn best. If we are successful in this we may be able to help instructors design their courses with student needs and competencies in mind. This instrument could also help counselors advise students as to what courses to take and how to study for them.

Answer TRUE or FALSE to each statement in the questionnaire. If a particular statement applies to you, check TRUE. If a particular statement does not apply to you, check FALSE. In answering each question, try to think in terms of how you go about learning in general, rather than thinking of a specific course or subject area. Be accurate and honest in your answers. Be sure to complete all the items, but do not spend a great deal of time on any one of them. This survey is for research use only and all information is kept confidential.

	TRUE	FALSE
1. When studying for an exam, I prepare a list of probable questions and answers.	✓	
2. I have trouble making inferences.		✓
3. I increase my vocabulary by building lists of new terms.	✓	
4. I am very good at learning formulas, names and dates.		✓
5. New concepts usually make me think of many other similar concepts.	✓	
6. Even when I feel that I have learned the material, I continue to study.	✓	
7. I have trouble organizing the information that I remember.		✓
8. Even when I know I have carefully learned the material, I have trouble remembering it for an exam.		✓
9. After reading a unit of material, I sit and think about it.	✓	
10. I make simple charts and diagrams to help me remember material.	✓	
11. I generally write an outline of the material I read.		✓
12. I try to convert facts into "rules of thumb".	✓	
13. I do well on tests requiring definitions.	✓	
14. I usually refer to several sources in order to understand a concept.	✓	
15. I ignore conflicts between the information obtained from different sources.		✓
16. I spend more time studying than most of my friends.	✓	

17. I learn new words or ideas by visualizing a situation in which they occur.   ____ ____

18. I learn new concepts by expressing them in my own words.   ____ ____

19. I often memorize material that I do not understand.   ____ ____

20. For examinations, I memorize the material as given in the text or class notes.   ____ ____

21. I carefully complete all course assignments.   ____ ____

22. I have difficulty planning work when confronted with a complex task.   ____ ____

23. I remember new words and ideas by associating them with words and ideas I already know.   ____ ____

24. I review course material periodically during the quarter.   ____ ____

25. I often have difficulty finding the right words for expressing my ideas.   ____ ____

26. Toward the end of a course, I prepare an overview of all material covered.   ____ ____

27. I find it difficult to handle questions requiring comparison of different concepts.   ____ ____

28. I generally read beyond what is assigned in class.   ____ ____

29. I have difficulty learning how to study for a course.   ____ ____

30. I have a regular place to study.   ____ ____

31. I read critically.   ____ ____

32. I "daydream" about things I have studied.   ____ ____

33. I do well on completion items.   ____ ____

34. I make frequent use of a dictionary.   ____ ____

35. I learn new ideas by relating them to similar ideas.   ____ ____

36. When learning a unit of material, I usually summarize it in my own words.   ____ ____

37. I maintain a daily schedule of study hours.   ____ ____

38. I think.   ____ ____

39. While learning new concepts their practical applications often come to my mind.   ____ ____

40. I get good grades on term papers.   ____ ____

41. Getting myself to begin studying is usually difficult. ____ ____

42. When necessary, I can easily locate particular passages in a textbook. ____ ____

43. I can usually formulate a good guess even when I do not know the answer. ____ ____

44. I have trouble remembering definitions. ____ ____

45. I would rather read the original article than a summary of an article. ____ ____

46. While studying, I attempt to find answers to questions I have in mind. ____ ____

47. I can usually state the underlying message of films and readings. ____ ____

48. I work through practice exercises and sample problems. ____ ____

49. I find it difficult to handle questions requiring critical evaluation. ____ ____

50. I have regular weekly review periods. ____ ____

51. I do well on examinations requiring much factual information. ____ ____

52. Most of my instructors lecture too fast. ____ ____

53. I look for reasons behind the facts. ____ ____

54. I cram for exams. ____ ____

55. When I study something, I devise a system for recalling it later. ____ ____

56. I have trouble seeing the difference between apparently similar ideas. ____ ____

57. I always make a special effort to get all the details. ____ ____

58. I prepare a set of notes integrating the information from all sources in a course. ____ ____

59. My memory is actually pretty poor. ____ ____

60. I am usually able to design procedures for solving problems. ____ ____

61. I do well on essay tests. ____ ____

62. I frequently use the library. ____ ____

___
*Inventory of Learning Processes (Schmeck, 1983), by permission of the author.

Scoring Sheet for Inventory of Learning Processes

Give yourself one point for each statement that is TRUE or FALSE as indicated. For instance, if you answered FALSE to item #2, give yourself 1 point. If you answered TRUE, give yourself 0. When you have completed each of the four scales, add them up for a total score.

After you have scored each scale, double check your scoring. Finally, add each column for a total score for each scale.

SCALE 1:		SCALE 2:		SCALE 3:		SCALE 4:	
DEEP PROCESSING		ELABORATIVE PROCESSING		FACT RETENTION		METHODICAL STUDY	
2. FALSE	____	5. TRUE	____	4. TRUE	____	1. TRUE	
7. FALSE	____	9. TRUE	____	13. TRUE	____	3. TRUE	____
8. FALSE	____	12. TRUE	____	20. TRUE	____	6. TRUE	____
15. FALSE	____	17. TRUE	____	33. TRUE	____	10. TRUE	____
19. FALSE	____	18. TRUE	____	44. FALSE	____	11. TRUE	____
22. FALSE	____	23. TRUE	____	51. TRUE	____	14. TRUE	____
25. FALSE	____	32. TRUE	____	59. FALSE	____	16. TRUE	____
27. FALSE	____	35. TRUE	____			21. TRUE	____
29. FALSE	____	36. TRUE	____			24. TRUE	____
31. TRUE	____	39. TRUE	____			26. TRUE	____
38. TRUE	____	46. TRUE	____			28. TRUE	____
40. TRUE	____	53. TRUE	____			30. TRUE	____
43. TRUE	____	55. TRUE	____			34. TRUE	____
47. TRUE	____	60. TRUE	____			37. TRUE	____
49. FALSE	____					41. FALSE	____
52. FALSE	____					42. TRUE	____
56. FALSE	____					45. TRUE	____
61. TRUE	____					48. TRUE	____
						50. TRUE	____
						54. FALSE	____
						57. TRUE	____
						58. TRUE	____
						62. TRUE	____

_____		_____		_____		_____	
TOTAL SCALE 1		TOTAL SCALE 2		TOTAL SCALE 3		TOTAL SCALE 4	

OVERALL GRADE POINT AVERAGE: _____

GENDER: MALE ____   FEMALE ____

AGE: _____

Turn this into your instructor when it is completed. Do not put your name on it.

LABORATORY EXERCISE 2: RECALL OF DREAMS

The purpose of Laboratory Exercise 2 is to demonstrate how dream researchers collect dream reports from subjects at home and how dream recall can be assisted by following a few simple suggestions. In Laboratory Exercise 3 you will do a content analysis of your dreams.

For the night your instructor assigns, you are to write down in detail all the dreams you recall during that night. Bring your description of your dreams to the assigned class period for discussion.

To assist yourself in dream recall, do the following:

1. Tell yourself before you go to sleep: "I shall remember all of my dreams tonight. It will be easy to write them down during the night or in the morning when I wake up."

2. Have paper and pencil next to your bed so that you can easily record your dreams.

3. When you wake up, do not move. Instead, review your dreams while you remain in your drowsy state. After you have reviewed your dreams, write them down.

4. If you cannot remember your dreams, set your alarm clock to go off 90 minutes after you go to sleep. This should wake you up during your first dream cycle.

For each of the dreams you recall, do the following:

1. Write each dream in as much detail as possible, in present or past tense.

2. If there are parts of the dream that you do not remember, just indicate that there are parts not remembered.

3. Describe the emotions and feelings that were in the dream, both your own and others.

4. After you have completed the description, read the dream over slowly and determine if you can remember anything else. If so, record it.

5. On a separate sheet of paper, write down how you react to this dream.

6. Finally, write down what you think is (are) the meaning(s) of the dream. Is it related to anything you did or said in the past few days? Is it related to any events in the past? Is it about any upcoming events in the future?

7. When you have finished 1-6 for each of your dreams, on a separate page, describe any connections you think may exist between your dreams.

Participant observation can be biased. Think about how you may have only partially recalled the dreams or actively censured your written report.

LABORATORY EXERCISE 3: CONTENT ANALYSIS OF DREAMS

The purpose of Laboratory Exercise 3 is to give you some experience in analyzing the content of dreams.

Dream researchers have studied the content of home and sleep laboratory dream reports from college students (e.g., Hall & Van de Castle, 1966b; Hall, Domhoff, Blick, & Weesner, 1982) and children (Foulkes, 1981). The content of "laboratory" and "home" dreams are sometimes different (e.g., Domhoff & Kamiya, 1964; Hall & Van de Castle, 1966). This is not surprising since we may incorporate the dream lab and its people into our lab dreams. When reporting our dreams to others, we may also actively censure our reports.

Hall et al. (1982) posed the question whether college students of today dream about the same things as their counterparts 30 years ago when Hall and Van de Castle (1966b) asked students to keep dream diaries. They used some of the Hall-Van de Castle (1966a) scales of dream content: "characters; aggressive, friendly, and sexual interactions; misfortunes; settings; and two types of objects, clothing and weapons" (p. 189).

They found that dreams of middle- and upper-middle class college students in 1950 and 1980 were quite similar in content. Somewhat surprisingly, the gender differences in the 1980 sample are the same as those in the 1950 sample. In Table 1, on the next page, some of the proportions for their dream categories are presented. A proportion was determined by dividing the top number of the formula by the bottom one. For instance, to determine the proportion of male characters in a dream one would take the total number of male characters and divide that by the total number of both male and female characters (Males/Males + Females). If there were an equal number of men and women it would be .50; if there were more men than women, it would be greater than .50. Now look at Table 1. You will find that men in both samples have proportions of .67; this means that the male subjects report more male than female characters. By contrast, the female subjects report significantly fewer male characters. In the Table the subscripts a and b show at what probability level (.001 or .01) the studies found significant gender differences. You should now be able to interpret the other categories of dream content.

For one or more of your dreams (as assigned by the instructor), you are to do a content analysis. This means that you are to read through your description and determine for each dream: (1) the number of male and female characters that were present in your dream; (2) the presence and number of verbal and physical aggressions there were; (3) the number of male and female characters who were doing the aggression; (4) whether there was some sort of sexual activity present; (5) whether misfortunes were present; (6) whether the settings were inside or outside; and (7) whether weapons were present. Use the worksheet provided at the end of this laboratory exercise to record your tallies. When this is completed, develop proportions for each of the categories, following the formulas which are presented at the bottom of the worksheet.

Are your dream contents similar to the norms in Table 1?

As a further class project, you may wish to determine the average proportions of all your classmates and compare this average with those presented in Table 1.

Sleep References:

Antrobus, J. (1991). Dreaming: Cognitive processes during cortical activation and high afferent thresholds. *Psychological Review, 98*, 96-121.

Antrobus, J. S., & Bertini, M. (Eds.) (1992). *Neuropsychology of sleep and dreaming*. Hillsdale, NJ: Lawrence Erlbaum Associates, Inc.

Cabel, S. (1989). Dreams as a possible reflection of a dissociated self-monitoring system. *Journal of Nervous and Mental Disease, 177*, 560-568.

Cartwright, R. L., & Kasniak, A. (1978). The social psychology of dream reporting. In A. M. Arkin, J. S. Antrobus, & S. J. Ellman (Eds.), *The mind in sleep*. Hillsdale, NJ: Lawrence Erlbaum.

Cipolli, C., Baroncini, P., Fagioli, I., Fumai, A., et al. (1987). The thematic continuity of mental sleep experience in the same night. *Sleep, 10*, 473-479.

Domhoff, B., & Kamiya, J. (1964). Problems in dream content with objective indicators: I. A comparison of home and laboratory reports. *Archives of General Psychiatry, 11*, 519-524.

Fitch, T., & Armitage, R. (1989). Variations in cognitive style among high and low frequency dream recallers. *Personality and Individual Differences, 10*, 869-875.

Foulkes, D. (1979). Home and laboratory dreams: Four empirical studies and a conceptual reevaluation. *Sleep, 2*, 233-251.

Foulkes, D. (1981). *Children's dreams*. New York: Wiley.

Foulkes, W. D. (1985). *Dreaming: A cognitive-psychological analysis*. Hillsdale, NJ: Lawrence Erlbaum Associates.

Hall, C. S. (1984). "A ubiquitous sex difference in dreams" revisited. *Journal of Personality and Social Psychology, 46*, 1109-1117.

Hall, C. S., Domhoff, W., Blick, K. A., & Weesner, K. E. (1982). The dreams of college men and women in 1950 and 1980: A comparison of dream contents and sex differences. *Sleep, 5*, 188-194.

Hall, C. S., & Van de Castle, R. L. (1966a). *The content analysis of dreams*. New York: Appleton-Century-Crofts.

Hall, C. S., & Van de Castle, R. L. (1966b). Studies of dreams collected in the laboratory and at home. *Institute of Dream Research Monograph Series*, No. 1. Felton, CA: Big Tree Press.

Hobson, J. A. (1988). *The dreaming brain.*. New York: Basic Books.

Lortie-Lussier, M.,. Simond, S., Rinfred, N., & de Koninck, J. (1992). Beyond sex differences: Family and occupational roles' impact on women's and men's dreams. *Sex Roles, 26*, 79-96

Mamelak, A. N., & Hobson, J. A. (1989). Dream bizarreness as the cognitive correlate of altered neuronal behavior in REM sleep. *Journal of Cognitive Neuroscience, 1*, 201-222.

Weinstein, L., Schwartz, D. H., & Ellman, S. J. (1988). The development of scales to measure the experience of self-participation in sleep. *Sleep, 11*, 437-447.

Wood, J. M., Bootzin, R. R., Rosenhan, D., Nolen-Hoeksema, S., & Jourden, F. (1992). Effects of the 1989 San Francisco earthquake on frequency and content of nightmares. *Journal of Abnormal Psychology, 101*, 219-224.

TABLE 1

Comparisons of 1950 Men and Women with 1980 Men and Women*

Content Variable	1950 STUDENTS		1980 STUDENTS	
	Men	Women	Men	Women
Characters				
1. Males/Males + Females	0.67	0.48[a]	0.67	0.53[a]
2. Familiar/Familiar + Unfamiliar	0.45	0.58[a]	0.55	0.62[a]
Aggression				
1. Aggression/Characters	0.34	0.24[a]	0.32	0.23[b]
2. Aggression with males/Males	0.28	0.22[a]	0.30	0.23[b]
3. Aggression with females/Females	0.17	0.14	0.13	0.15
4. Physical aggression/Physical + verbal aggression	0.50	0.34[a]	0.57	0.39[a]
Friendliness				
1. Friendliness/Characters	0.21	0.22	0.16	0.17
2. Friendliness with Males/Males	0.17	0.24[a]	0.09	0.20[a]
3. Friendliness with Females/ Females	0.29	0.15[a]	0.28	0.15[a]
Sex, misfortunes, settings, & objects				
1. Dreamers with at least one sex/ No. dreamers	0.38	0.17[a]	0.24	0.10[b]
2. Misfortunes/No. dreams	0.41	0.41	0.36	0.43
3. Outdoor settings/Outdoor + Indoor	0.51	0.39[a]	0.49	0.37[a]
4. Clothes/No. dreams	0.28	0.54[a]	0.10	0.34[a]
5. Weapons/No. dreams	0.12	0.03[a]	0.15	0.04[a]

a $p < 0.001$, b $p < 0.01$

*Adapted from Hall et al. (1982, p. 192, Table 2) with permission of Raven Press.

Content Analysis of Your Dreams

Category	Dream 1	Dream 2	Dream 3	Proportion*
Characters:				
Number of Males				
Number of Females				
Total				
Aggression: Type				
Number of verbal aggressions				
Number of physical aggressions				
Total				
Aggression: number of				
Males involved				
Females involved				
Total				
Sexual activity of some sort present				
Misfortunes present				
Settings:				
Indoor				
Outdoor				
Total				
Weapons Present				

* To determine proportions:

1) # of males/males + females
2) # of females/males + females
3) Aggressions/# of dreams
4) Sexual activity/# of dreams
5) Misfortunes/# of dreams
6) Indoor/indoor + outdoor
7) Outdoor/indoor + outdoor
8) # of dreams with weapons/# of dreams

TOPIC 2: PROBLEM IDENTIFICATION AND HYPOTHESIS TESTING

The goals of Topic 2 are to provide you with instruction in conducting a literature review and to give you practice in conducting a limited literature review. Additionally, you will learn how to formulate a research question and hypothesis based upon this review.

When preparing to do research, scientists should conduct a thorough review of the pertinent literature contained in journal articles, books, and dissertations. While the sources of our ideas for psychological experiments come from many places, we need to know what is and is not known about the research area. The literature review helps to sharpen the experimental question so that it can be transformed into a testable experimental hypothesis.

The body of the psychological literature is enormous, containing a tremendous number of journals and books. Fortunately there are indexes available which have summarized and categorized much of this literature. When conducting a literature review, psychologists often find the most valuable index sources to be *Psychological Abstracts, Index Medicus, Social Sciences Citation Index,* and *Science Citation Index*. The introduction of computerized literature searches saves you a tremendous amount of time. An introduction to each of these indexes is provided below.

PSYCHOLOGICAL ABSTRACTS

Psychological Abstracts is issued monthly in journal form by the American Psychological Association. It contains brief abstracts of studies in all areas of psychology from almost 1000 journals. The *Abstracts* has been the primary source of information about psychological literature since 1927. An earlier publication, the *Psychological Index*, covers literature published between 1894 and 1935. In order to know where to look in the *Psychological Abstracts*, since the literature is categorized under many headings, it is important to use the *Thesaurus of Psychological Index Terms*, which is also published by the American Psychological Association to accompany *Psychological Abstracts*.

The *Thesaurus* will help you identify terms and headings related to your research topic. "Each *Thesaurus* is listed alphebetically and, as appropriate, is cross-referenced and displayed with its broader, narrower, and related terms (i.e., subterms)" (APA, 1994, p. vii). The subject code (SC) may be used instead of the term. The posting note (PN) indicatels how many times the term has been used. The scope notes (SN) define the term. In addition, broader (B), narrower (N), and related (R) terms are provided.

If you were interested in dreaming, two general topic areas that probably came to your mind immediately are dreams and sleep. In the seventh edition of the *Thesaurus of Psychological Index Terms* (1994), the relationship section[1] provides many headings that are related to dreaming and sleep. Some relevant headings are provided on the next page.

---

[1]*Thesaurus of Psychological Index Terms* (7th ed.), 1994, pp. 64 and 203. Copyright 1994 by the American Psychological Association. Reprinted by permission.

Dream Analysis 73
PN 762   SC 15060
  UF   Dream Interpretation
  B   Psychoanalysis 67
       Psychotherapeutic Techniques 67
  R   ↓ Dreaming 67
      ↓ Parapsychology 67

Dream Content 73
PN 757   SC 15070
  R   ↓ Dreaming 67
      Nightmares 73
      ↓ Sleep 67

Dream Interpretation
  Use   Dream Analysis

Dream Recall 73
PN 215   SC 15090
  R   ↓ Dreaming 67
      Lucid Dreaming 94

Dreaming 67
PN 1109   SC 15100
  N   Lucid Dreaming 94
      Nightmares 73
      REM Dreams 73
  R   Dream Analysis 73
      Dream Content 73
      Dream Recall 73
      ↓ Sleep 67

Sleep 67
PN 4159   SC 47820
  N   Napping 94
      NREM Sleep 73
      REM Sleep 73
  R   ↓ Consciousness Disturbances 73
      ↓ Consciousness States 71
      Dream Content 73

Sleep — (cont'd)
  R   ↓ Dreaming 67
      Lucid Dreaming 94
      Nocturnal Teeth Grinding 73
      Sleep Apnea 91
      Sleep Deprivation 67
      ↓ Sleep Disorders 73
      Sleep Onset 73
      Sleep Talking 73
      Sleep Treatment 73
      Sleep Wake Cycle 85

Sleep Apnea 91
PN 31   SC 47825
SN Temporary absence of breathing or prolonged respiratory failure occurring during sleep.
  B   Apnea 73
  R   ↓ Neonatal Disorders 73
      ↓ Sleep 67
      Sudden Infant Death 82

Sleep Deprivation 67
PN 810   SC 47830
  B   Deprivation 67
  R   ↓ Sleep 67
      ↓ Sleep Disorders 73

Sleep Disorders 73
PN 822   SC 47840
  UF   Night Terrors
  B   Consciousness Disturbances 73
      Disorders 67
  N   Hypersomnia 94
      Insomnia 73
      Narcolepsy 73
      Sleepwalking 73
  R   Hypnagogic Hallucinations 73
      ↓ Sleep 67
      Sleep Deprivation 67

Sleep Talking 73
PN 10   SC 47870
  B   Consciousness Disturbances 73
  R   ↓ Sleep 67

Once you have determined which subject headings may be helpful to you in your literature review, you can turn to the *Psychological Abstracts*. In the back of each monthly issue is an author and subject index. An abbreviated example[2] of a typical index to a month issue of the *Psychological Abstracts* is shown below.

Following each heading are a list of numbers. They refer to the number (not page) of individual abstracts in that issue. If we were interested in "Dream Recall" we would look up 357[3], which is the following:

> 357. **Fitch, Thomas & Armitage, Roseanne.** (Carleton U. Ottawa, ON, Canada) **Variations in cognitive style among high and low frequency dream recallers.** *Personality & Individual Differences,* 1989, Vol 10(8), 869–875. —Evaluated creative thinking among 17 male high frequency dream recallers (HFDRs), 14 female HFDRs, 20 male low frequency dream recallers (LFDRs), and 19 female LFDRs (aged 17–40 yrs) identified by a frequency of dream recall questionnaire. Ss also completed the Torrance Tests of Creative Thinking. HFDRs generated more elaborate and original stimulus uses than LFDRs. HFDRs, and in particular females, were more fluent in processing picture completion tasks. Results reflected higher cognitive flexibility and more divergent information processing among HFDRs. The ability to recall dreams is a fundamental cognitive style variation in sleep and in wakefulness. The frequency of dream recall questionnaire is appended.

This abstract provides much information. It tells you who authored the paper and where they are located (or where they were located when the research was carried out), the title of the paper, and where and when it was published. The abstract gives you a good summary of the article. There is a general statement of its purpose, how it was carried out, what the major results were, and often a theoretical interpretation of the major findings.

The above method is to be followed for the very latest issues of *Psychological Abstracts*. If you wish to search literature which is older, there is an easier way than searching each month separately. For every year through 1983 there are two volumes of the *Abstracts* issues with a more indepth index to each issued volume. Beginning with 1984 there is one volume with a Cumulated Annual Index. Brief subject and author indexes cumulate quarterly (i.e., March index at the end of the monthly issue includes abstracts in January, February, and March issues). Compare the following material from the 1982 (July-December) semiannual index[4] under the headings of "Dream Content", "Dream Recall", and "Dreaming" to that given before.

Dream Content
classification-rating system for anomalies in dreams, 9623
dream content & sex differences, college students, 1950 vs 1980, 11930
failure of defensive functions of ego in dreams, psychosomatic patients, 8484
manifest dream content & object relations, level of interpersonal maturity & mental health, 21–39 yr old neurotic vs character disordered females, 1274
mental vs cultural vs organismic vs physical correlates, dream content, 15–30 yr olds, 12387
performance of certain tasks during lucid-dream state, 15–71 yr old lucid dreamers, 7793
repetitive dream content & aggressive & sexual drive & role of ego, adolescent & young adult female anorexia nervosa patients during bulimic phase, 1499
validity of Initial Letter Word Association Test, prediction of form of dream contents, 9472
war environment, number & content of dreams & sleep disturbances & repression sensitization, 14–17 yr olds, Israel, 10040
Dream Interpretation [See Dream Analysis]
Dream Recall
extraversion & anxiety & belief in ESP & dream recall & imagery, high vs low scorers in psi task, 2324

Dreaming [See Also Nightmares, REM Dreams]
classification-rating system for anomalies in dreams, 9623
dream healings through hypnosis, ancient Greek Asclepian temples, literature review, 6391
dream structure & dynamics, unabridged reprint of essay by S. De Sanctis, 9624
Freud's model of verbal behavior in dreams, 2778
"light switch phenomenon" in lucid dreams, male & female lucid dreamers, 11931
parapsychological phenomena during dreams & other altered consciousness states, 2317
performance of certain tasks during lucid-dream state, 15–71 yr old lucid dreamers, 7793
psycholinguistic model of speech production as basis for model of dream production, 11926
war environment, number & content of dreams & sleep disturbances & repression sensitization, 14–17 yr olds, Israel, 10040

---

[3]*Psychological Abstracts, 77,* 1990, p. 35. Copyright 1990 by the American Psychological Association. Reprinted by permission.

[4]*Psychological Abstracts, 68,* 1982, p. 302. Copyright 1982 by the American Psychological Association. Reprinted by permission.

More information is given about each article with key words to help you decide whether the article may be of interest to you. For instance, the entry "dream content & sex differences, college students, 1950 vs 1980, 11930" sounds very close to the content analysis laboratory exercise we did in Topic 1. These few words tell us that the study involves the comparison of dreams between college students in 1950 and 1980 and the study of sex differences. Below is Abstract #11930[5], corresponding to this entry.

> 11930. Hall, Calvin S.; Domhoff, G. William; Blick, Kenneth A. & Weesner, Kathryn E. (U California, Santa Cruz) **The dreams of college men and women in 1950 and 1980: A comparison of dream contents and sex differences.** *Sleep*, 1982, Vol 5(2), 188–194. —Two samples of dreams collected from 200 college students in 1950 and 122 students in 1980 were analyzed with the dream content scales devised by C. S. Hall and R. L. Van de Castle (1966). Items of comparison included dream characters, aggression, friendliness, sex, misfortune, and dream settings. Little change in what Ss dreamed about was found over this 30-yr period. Notably, the sex differences in the 1980 dreams were the same as those found in the 1950 dreams. Findings are discussed in light of the continuity and the compensatory hypotheses. (12 ref) —*Journal abstract.*

If you are carrying out literature searches which extend over many years, you should use the *Cumulative Subject Index to Psychological Abstracts*, 1927-1960. There are supplementary volumes for later years (1961-1965; 1966-68; 1969-71; and so on.) There is also the *Author Index to Psychological Index* (1894-1935) and *Author Index to Psychological Abstracts* (1927-1960), as well as more recent supplements.

One question frequently asked by students is "How far back must I search?" The answer depends upon how extensive of a search you need to do. Ideally, you should go back as far as there is relevant material.

CUMULATED INDEX MEDICUS

The National Library of Medicine publishes a *Cumulated Index Medicus* and monthly *Index Medicus* in much the same manner as *Psychological Abstracts*. It is a comprehensive index of the world's medical literature which has author and subject indexes. Unlike *Psychological Abstracts, Index Medicus* only gives the authors, the title, and where the article was published. It does not provide a summary of the article. Below is an exerpt[6] taken from the section on "DREAMS".

DREAMS

Sleep disturbances in torture survivors. Astrøm C, et al. Acta Neurol Scand 1989 Feb;79(2):150-4

Nightmares. Pagel JF Jr. Am Fam Physician 1989 Mar; 39(3):145-8 (13 ref.)

The manifest content of dreams. Gallego-Mere A. Am J Psychoanal 1989 Jun;49(2):95-103

Life-long nightmares: an eclectic treatment approach. Gorton GE. Am J Psychother 1988 Oct;42(4):610-8

Propofol and dreaming [letter] Gunawardene RD. Anaesthesia 1989 Mar;44(3):266-7

Selective recall of surprising visual scenes. An experimental note on Seligman and Yellen's theory of dreams. van den Hout MA, et al. Behav Res Ther 1989;27(2):199-201

Dreams in head-injured patients. Benyakar M, et al. Brain Inj 1988 Oct-Dec;2(4):351-6

The place of the dream in psychotherapy. Kaplan DM. Bull Menninger Clin 1989 Jan;53(1):1-17

A systematized method for dream analysis in a group setting. Shuttleworth-Jordan AB, et al. Int J Group Psychother 1988 Oct;38(4):473-89

The dream as religion of the mind. Mancia M. Int J Psychoanal 1988;69 (Pt 3):419-26

The thematic continuity of mental experiences in REM and NREM sleep. Cipolli C, et al. Int J Psychophysiol 1988 Nov;6(4):307-13

Life attitudes, dreams, and psychic trauma in a group of "normal" children. Terr LC. J Am Acad Child Psychiatry 1983 May;22(3):221-30

SCIENCE CITATION INDEX AND SOCIAL SCIENCE CITATION INDEX

Published by the Institute for Scientific Information, the *Science Citation Index* is an interdisciplinary work covering articles in science, medicine, agriculture, technology and the behavioral sciences. Its companion series, the *Social Science Citation Index*, covers psychological literature as well as other social science areas. There are three section in each:

A. <u>Permuterm Subject Index</u>. This index should be used first in any literature search as it gives the significant terms for the titles of every article covered in the given index period. In other words, it pairs every two significant terms from each title written. If you are interested in state-dependent learning, you would look up "learning" and scan the words listed under it for "state-dependent". The names of authors who have published articles with these two terms in their titles will be listed. If your library has the Citation Indexes on magnetic tape, your research is made more effortless. All you need to do is punch in the significant terms, "learning" and "state-dependent" in this instance, and the computer will print out a bibliography.

B. <u>Source Index</u>. This index gives you full bibliographic information by author, for all articles and books written and indexed during the year currently covered. You are given the author's mailing address, the complete reference, and its complete referencing, as well as the number of references there are in each article.

C. <u>Citation Index</u>. This Index lists all authors whose previous words were cited during the year covered. Every article that has made reference to an author's earlier article will be listed.

A separate, but related work is the *Index to Scientific Reviews*. This is an excellent place to start a review of the literature as it indexes solely the comprehensive review articles.

COMPUTERIZED LITERATURE SEARCHES

Computerized literature searches are now available for researchers and scholars in a number of fields. Psychologists are most apt to use the following information retrieval systems: *PsycLIT, Medline, ERIC* (educational psychology is covered here), and the *Social Science Citation Index*. Your library may have one or more of these systems, and a reference librarian can help you learn how to use them. Librarians will do a recovery search service using one or more of these retrieval systems for a fee. Recently, libraries have begun to purchase CD-ROM disks that can be used on personal computers for either a minimal or no cost. For instance, PsycLIT and Medline are available since 1974 and are updated quarterly.

The major advantage of computerized literature searches is speed. A keyword search accesses the computer's memory, scanning and pulling up all titles and abstracts of articles that have the keyword(s) within them. For the most efficient and indepth search, you will probably use two or more keywords. The accompanying thesaurus is of major assistance to your determining appropriate keywords. That computerized searches are limited to the choice of keywords put in can be considered a potential disadvantage.

On the following pages are instructions as to how to use PsycLIT. Note the library handout instructions[7] on how to select keywords to find articles on the effect of television violence on children (one of our own later laboratory exercises). Once the computer chooses the pertinent articles, you are informed how many were found. You can then view each reference, with or without the accompanying abstract. Often you can either download the references to your own floppy disk for subsequent personal[8] use or print out the information to an online printer.

---

[7]Library handout reproduced with permission of the Virginia Polytechnic Institute and State University library. PsycLIT citations reprinted with the permission of the American Psychological Association, publisher of PsycLIT and the PsycINFO data base (copyright 1967-1993 by the American Psychological Association). May not be reproduced without its prior permission.

[8]All information retrieval systems are copyrighted by the individual publisher and any redistribution of the data other than for personal use must be by permission of the publisher.

PsycLit

The PsycLIT CD-ROM database contains summaries of the world's serial literature in psychology and related disciplines, covering over 1300 journals, and corresponds to the print index Psychological Abstracts [BF1 P65 Ref Room]. Coverage is from 1974 - present, with quarterly updates. The PsycLIT database is on two discs: 1974-1982 and 1983-present.

1 BASIC SEARCHING

A. Press any key to see the Library's CD-ROM network menu and then type **B** to choose PsycLIT. Use the up and down arrow keys to select the disc to be searched (current disc is highlighted). Press **ENTER.** Wait for the **FIND** prompt to appear.

B. Type the words and/or phrases relevant to your topic. The Thesaurus of Psychological Index Terms or the on-disc thesaurus (**F9**) will be helpful in determining appropriate search terms.

C. Enter each search term or phrase separately, using hyphens between multiple word descriptors and after single word descriptors.
Examples: drug-abuse
              hypnotherapy-
(If you use the on-disc thesaurus, you may select terms to be searched automatically, without typing.) The number of records retrieved as each term is searched will display as separately numbered sets. You may look at any set using the SHOW command (**F4**), or you may refine the search by adding more terms.

D. **Free-text terms:** Free-text terms are words or phrases from several different fields, including title, abstract, descriptor, and key phrase. Sometimes there are not pertinent descriptor terms listed in the thesaurus, in which case simply enter all terms you think relevant to the topic.

2 REFINING YOUR SEARCH

       AND   OR   IN

Use the above connectors to combine sets to broaden or narrow a search:

Examples:
violence OR crime
Retrieves records containing one or the other term or both terms. **OR** is used to broaden a search.

violence AND crime
Retrieves records containing both words. **AND** is used to narrow a search.

violence IN AB
Retrieves records containing a word or phrase within a specified field (in this case, the **abstract** field).

Additional connectors are **NOT, NEAR** and **WITH.** For help with these, consult the SilverPlatter manual or ask for assistance.

3 DISPLAYING YOUR RESULTS

A. To look at the records, press **F4** or **SHOW** from the command menu (to get to the command menu press **F10** or **ESC**).

B. To "mark" records to be printed or downloaded, press **ENTER** or **M** with the cursor on any line in the record.

C. Use **CTRL-[PGDN]** to move to the top of the next record, or **CTRL-[PGUP]** to move to the top of the previous record.

D. To select additional search terms from retrieved records place the cursor on the desired term and type **S** from the **SHOW** command menu.

4 PRINTING YOUR RESULTS

A. Press **F6** or **P** from the **SHOW** command menu.

B. The system defaults to CITN (citation) for printing. Press **ENTER** to accept default settings or type **C** from the **PRINT** command menu to specify the information to be printed. Use **TAB** key to select options to be changed and **SPACE** bar to change settings. Press **ENTER** to accept changes. Press **ENTER** again to start printing.

5 DOWNLOADING YOUR RESULTS

A. Place your formatted disc in a disc drive. (Disc drives are labeled A, B, or D.)

B. Press **D** from the **SHOW** command menu or **D** from the command menu (**F10** or **ESC**). The Downloading Options Screen will appear. Press ENTER to accept default settings, or type **C** to change options. Use **TAB** key to select option to be changed and **SPACE** bar to change settings. Press **ENTER** to accept changes. Press **ENTER** again to download.

6 HINTS

Truncating: The * is the truncation symbol and can be used to retrieve all variations of the root term. Example: *FIND obsess** retrieves obsess, obsessed, obsessive, obsessing, obsessional.

Displaying Records: Ctrl-[PgUp] & Ctrl-[PgDn]

Fields: Press **F3** to see a list of fields included in a PsycLIT record.

Printing Limits: The Library has imposed a limit on the number of records that can be printed, so choosing ALL will not print all records if that number exceeds our limit.

Set #s: You can only display, print or download results from the most recent set. To view an earlier set, type # followed by the number of the desired set (example: #4). The set is reselected and renumbered.

7 USING FUNCTION KEYS

F1	HELP	Explains commands and other features of database
F2	FIND	Searches for words or phrases in the database
F3	GUIDE	Acts as a handbook to PsycLIT database
F4	SHOW	Displays retrieved records on screen; allows selection of additional words and/or phrases to be searched. Records can be "marked" for printing or downloading.
F5	INDEX	Alphabetically lists indexed words in database; allows selection of entries for searching
F6	PRINT	Prints retrieved records or "marked" records. Print options can be changed from print command window. To change options type C.
F7	RESTART	Begins or ends a session; returns to intial screen; search history will be cleared.
F8	XCHANGE	Permits exchange of disc within PsycLIT database. Search history will be saved.

F9	THESAURUS	Use to determine appropriate search terms. Terms can be selected and searched from the Thesaurus.
F10	COMMAND MENU	The **ESC** key also causes this menu to appear

8 PERFORMING A SAMPLE SEARCH

RESEARCH QUESTION:
What is the effect of television violence or crime on children?

SAMPLE SEARCH

Type a search request, then press RETURN.
FIND: (TELEVISION or TV-) and (VIOLENCE- or CRIME-) and CHILDREN-

SilverPlatter 2.01		PsycLit Disc 2 (1/83-3/91)
No.	Records	Request
#1	314	TELEVISION-
#2	1222	TV-
#3	1243	TELEVISION- or TV-
#4	993	VIOLENCE-
#5	1917	CRIME-
#6	2742	VIOLENCE- or CRIME-
#7	4183	CHILDREN-
#8	5	#3 and #6 and #7

TO **SHOW** records found, press F4.

SAMPLE RECORD RETRIEVED

SilverPlatter 2.01   PsycLIT Disc 2 (1/83-3/91)

TI: Television as a source of maltreatment of children
AU: Eron,-Leonard-D.; Huesmann,-L.-Rowell
IN: U Illinois, Chicago
JN: School-Psychology-Review; 1987 Vol 16(2) 195-202
CO: SPDID7
IS: 02796015
LA: English
PY: 1987
AB: Discusses the effect of TV on children's behavior, especially on aggressive and prosocial behavior. It is maintained that heavy exposure to TV violence is one of the causes of aggressive behavior, crime, and violence in society. Research findings that have established relations between viewing and the subsequent attitudes and behaviors of viewers are reviewed. Implications and remedies for public policy, society, parents, and educators are identified. (PsycLIT Database Copyright 1987 American Psychological Assn, all rights reserved)
KP: TV viewing; social development & aggressive & prosocial behavior; children; implications for public policy & parental & societal responsibility
DE: TELEVISION-VIEWING; PSYCHOSOCIAL-DEVELOPMENT; AGGRESSIVE-BEHAVIOR; PROSOCIAL-BEHAVIOR; CHILDREN-; RESPONSIBILITY-; GOVERNMENT-POLICY-MAKING
CC: 2840
PO: Human
UD: 8710
AN: 74-27835

9 ENDING YOUR SEARCH
Press **F10** or **ESC** and type **Q**

10 ASKING FOR HELP
Library staff at the Electronic Reference Area desk are happy to assist you.

LABORATORY EXERCISE 1: SEARCHING THE LITERATURE

The purpose of this laboratory exercise is to give you experience in using *Psychological Abstracts* or *PsycLIT*. Refer back to the previous section for instructions as to how to use these literature search systems.

1. Identify a specific area of research in which you are interested. For instance, you might be interested in the effect of cocaine on cognitive functioning in humans. The more specific and clear you are (e.g., cocaine, rather than psychoactive drugs in general), the easier your search of the literature will be. You may be interested in general cognitive functioning, or something more specific such as memory or dreaming. Write your research question in the space below:

2. Go to the most recent edition of the *Thesaurus of Psychological Index Terms*, published by the American Psychological Association, or to the on-disc thesaurus of PsycLIT. In the relationship section, determine what headings best describe your research area. Write them below.

3. Once you have determined which subject headings may be helpful, look up the headings in an index of an individual issue or a semiannual index of *Psychological Abstracts*. Now look up the individual references that appear to be of interest to you. (If you are using *PsychLIT*, type in the heading(s) after "Find" and follow the instructions provided previously to examine the individual references and their abstracts.)

4. On separate paper, provide the following information about five articles you found to be of interest. Eliminate those articles that are written in a foreign language or are dissertations. Turn in this laboratory exercise sheet along with the following information you have obtained from the abstracts of each of the five articles.

Author(s):

Title:

Journal, year published, volume, and pages:

Purpose of article:

Specific hypothesis(es):

Summary of results:

LABORATORY EXERCISE 2: IDENTIFYING RESEARCH PROBLEMS AND HYPOTHESES

In Laboratory Exercise 1 you were to find five references pertinent to a research area you identified. The purpose of this laboratory exercise is to give you practice in identifying research problems and hypotheses that follow from them.

Read one research article from the five articles you identified in Laboratory Exercise 1. Be sure to choose a research article and not a review article.

As you read the article, identify the research problem and hypotheses of the study. A hypothesis is a prediction that follows logically from the statement of a problem. While the hypothesis is usually (but not always) explicitly stated, you may find that the statement of the problem is implicit and only determined from reading a review of the literature in the introduction.

We distinguish between the scientific hypothesis and the null hypothesis. The scientific hypothesis is the predicted effect or relationship among the variables. The null hypothesis is the statement that there will be no effect or relationship among the variables. In scientific research we test the null hypothesis with statistical tests. To provide support for the scientific hypothesis, we must obtain support for the rejection of the null hypothesis. You will learn, or have already learned, about statistical hypothesis-testing theory in your statistics courses.

Answer the following questions about the research article, and turn them into your instructor. Attach a copy of the article to your answers.

1.   Briefly describe the topic of the research.

2.   What is (are) the purpose(s) of the present research study? What is the problem that is under investigation?

3.   State the hypotheses of the present research study. For each hypothesis, give the scientific hypothesis and the null hypothesis.

Finally, based upon the research article you read, formulate one further hypothesis that logically follows from the research carried out. State both the scientific hypothesis and the null hypothesis.

TOPIC 3: OPERATIONAL DEFINITIONS

In order to do research and communicate meaningfully with others, we must define our concepts explicitly. First, you must develop a <u>conceptual definition of a concept</u> (examples of concepts would be frustration, love, anxiety, intelligence, etc.). When doing research, you must go beyond the conceptual definition and develop an explicit <u>operational definition of the concept</u> under investigation. This operational definition specifies the operations and measures which you use to define the concept. It must be presented in great enough detail so that another researcher could repeat it to replicate the study.

The term <u>operational definition</u> means that you are defining concepts by the operations used to attain them. For example, depression is defined conceptually as a state of being in which the individual exhibits lowered initiative and has sad and gloomy thoughts. Operationally, you could define the concept of depression in terms of behavioral observations (ex: affect level, content analysis of speech patterns), questionnaires (ex. Beck Depression Inventory), or physiological measures (ex: lateralization of EEG brain wave activity). In addition, you might wish to develop a depressing situation so as to try to induce a state of depression; if so, an operational definition of depression would be in terms of the procedure used to produce the emotion (ex: showing a very gloomy film to subjects).

The purpose of this Topic is to help you to understand the development of operational definitions of a concept. The concept of anxiety will be used in the first two laboratory exercises.

LABORATORY EXERCISE 1: DEFINING THE CONCEPT OF ANXIETY

Anxiety is an emotional state we have all experienced. Since the 1930s, after Freud emphasized its importance, both the psychological and medical fields have published thousands of studies investigating anxiety. They have studied procedures and stimuli which evoke anxiety and fear, as well as the responses which define the experience of anxiety, in both humans and non-humans.

As an exercise, pause for a few minutes and write down, in the space below, your definition of anxiety, as if you were writing a book of psychological terms.

Now, check the various definitions of anxiety given in a dictionary, either a general dictionary or a dictionary of psychological terms which can be obtained at the reference desk in your library. Write down those definitions of anxiety which you think would be useful to behavioral researchers.

Bring your own definition of anxiety and the definitions you obtained from a dictionary to class. Discuss your various conceptual definitions of anxiety. Choose one which you will use in the next laboratory exercise.

As you will probably note, anxiety is a much more complex concept than you may have thought it to be. As a result, the conceptual definition and theoretical orientation of a researcher can lead her/him down many different paths of research. The resulting operational definitions of different researchers may be similar or they may be quite different. The underlying question, of course, is whether the various operational definitions are measuring the same thing.

LABORATORY EXERCISE 2: OPERATIONALLY DEFINING THE CONCEPT OF ANXIETY

As a researcher it is not enough to have a conceptual definition of anxiety. You must now convert the conceptual definition to an operational definition.

Write the conceptual definition of anxiety which your class decided upon in the prior exercise.

Now, you are going to convert this conceptual definition into various kinds of operational definitions. To assist you in doing this you might think about ways in which you and others respond behaviorally and physiologically when experiencing anxiety, as well as how you feel internally.

List at least THREE ways by which you could measure anxiety at a physiological level. Be quite specific, remembering that your definition must be clear enough so that another researcher could replicate it.

Next, list at least THREE ways by which you can measure anxiety at a nonverbal, observational level. Specifically, what behaviors and/or conditions of an individual would indicate to an observer that she/he was experiencing anxiety?

List several ways by which you can measure anxiety through verbal reports. Specifically, what kind of questions would you ask to determine if an individual was experiencing anxiety? These can be true-false, multiple choice, or open-ended questions.

So far you have been operationally defining individuals' anxiety responses. Now list several ways by which an experimenter could experimentally produce anxiety with certain procedures and/or stimuli. Be sure that they are ethical. Once again be specific enough so that another researcher could replicate your procedure.

LABORATORY EXERCISE 3: SELF-REPORT QUESTIONNAIRES ASSESSING ANXIETY

In the prior laboratory exercise you listed a number of operational definitions for anxiety, including those that use self-reports from the individual. In this laboratory exercise you will fill out one self-report scale that has been used extensively to operationally define anxiety and then you will evaluate the psychometric properties of the scale. We shall also examine other scales which assess different types of anxiety.

One of the first anxiety self-report scales, the Manifest Anxiety Scale (MAS), was developed by Janet Taylor (1951, 1953). She is now Janet Taylor Spence, a well-known psychologist who was president of the American Psychological Association (1984-85).

The MAS is concerned with assessing your general characteristic anxiety level over time. The questions of the MAS are given below. Answer each statement true or false. Do not leave any blank. Be as honest as possible. When you complete the questionnaire, score yourself by using the scoring key which is below the MAS. Given yourself one point for each answer that is the same as that given.

Manifest Anxiety Scale

For each statement below check true or false as to how you generally feel.

	TRUE	FALSE
1. I do not tire quickly.	_____	_____
2. I am troubled by attacks of nausea.	_____	_____
3. I believe I am no more nervous than most others.	_____	_____
4. I have very few headaches.	_____	_____
5. I work under a great deal of tension.	_____	_____
6. I cannot keep my mind on one thing.	_____	_____
7. I worry over money and business.	_____	_____
8. I frequently notice my hand shakes when I try to do something.	_____	_____
9. I blush no more often than others.	_____	_____
10. I have diarrhea once a month or more.	_____	_____
11. I worry quite a bit over possible misfortunes.	_____	_____
12. I practically never blush.	_____	_____
13. I am often afraid that I am going to blush.	_____	_____
14. I have nightmares every few nights.	_____	_____
15. My hands and feet are usually warm enough.	_____	_____
16. I sweat very easily even on cool days.	_____	_____
17. Sometimes when embarrassed, I break out in a sweat which annoys me greatly.	_____	_____
18. I hardly ever notice my heart pounding and I am seldom short of breath.	_____	_____
19. I feel hungry almost all the time.	_____	_____
20. I am very seldom troubled by constipation.	_____	_____
21. I have a great deal of stomach trouble.	_____	_____
22. I have had periods in which I lost sleep over worry.	_____	_____
23. My sleep is fitful and disturbed.	_____	_____
24. I dream frequently about things that are best kept to myself.	_____	_____

		TRUE	FALSE

25. I am easily embarrassed. ____ ____
26. I am more sensitive than most other people. ____ ____
27. I frequently find myself worrying about something. ____ ____
28. I wish I could be as happy as others seem to be. ____ ____
29. I am usually calm and not easily upset. ____ ____
30. I cry easily. ____ ____
31. I feel anxiety about something or someone almost all the time. ____ ____
32. I am happy most of the time. ____ ____
33. It makes me nervous to have to wait. ____ ____
34. I have periods of such great restlessness than I cannot sit long in a chair. ____ ____
35. Sometimes I become so excited that I find it hard to get to sleep. ____ ____
36. I have sometimes felt that difficulties were piling so high that I could not overcome them. ____ ____
37. I must admit that I have at times been worried beyond reason over something that really did not matter. ____ ____
38. I have very few fears compared to my friends. ____ ____
39. I have been afraid of things or people that I know could not hurt me. ____ ____
40. I certainly feel useless at times. ____ ____
41. I find it hard to keep my mind on a task or job. ____ ____
42. I am usually self-conscious. ____ ____
43. I am inclined to take things hard. ____ ____
44. I am a high-strung person. ____ ____
45. Life is a strain for me much of the time. ____ ____
46. At times I think I am no good at all. ____ ____
47. I am certainly lacking in self-confidence. ____ ____
48. I sometimes feel that I am about to go to pieces. ____ ____
49. I shrink from facing a crisis or difficulty. ____ ____
50. I am entirely self-confident. ____ ____

SCORING KEY:

1 F	11 T	21 T	31 T	41 T
2 T	12 F	22 T	32 F	42 T
3 F	13 T	23 T	33 T	43 T
4 F	14 T	24 T	34 T	44 T
5 T	15 F	25 T	35 T	45 T
6 T	16 T	26 T	36 T	46 T
7 T	17 T	27 T	37 T	47 T
8 T	18 F	28 T	38 F	48 T
9 F	19 T	29 F	39 T	49 T
10 T	20 F	30 T	40 T	50 F

YOUR OWN TOTAL SCORE: _____ (Give to your instructor, on a separate sheet of paper, without your name.)

Using descriptive statistics, your class' scores can be summarized in several ways:

1. range: from the lowest to the highest score
2. mean: average score (sum of all scores divided by number of scores)
3. median: middle most score
4. mode: score with the most subjects (there may be more than one)

With the class' scores determine each of the above. Bar graph a distribution of these scores, labeling everything correctly.

Range: _____ to _____

Mean: _____   Median: _____   Mode: _____

Reflect for a moment: Did you like the true-false questions? Why? What other types of responses might you use to answer the same questions?

Turn this sheet into your instructor.

The MAS has been reevaluated psychometrically in a number of studies. The test-retest reliability (same subjects take it with a time period between to determine how consistent their scores are across time) is quite adequate: .89 over a three week period; .82 over a five-month period; .81 over a 9-17 month period (Taylor, 1953).

Attempts have been made to validate the MAS. When it is said that a test is valid, it means that the test really measures "anxiety" as it purports to measure it. For instance, the MAS correlates positively with other measures of anxiety: separation anxiety (Sarason, 1961), text anxiety (Sarason, 1961), and the belief that the environment is dangerous (Houston, Olson, & Botkin, 1972). It correlates moderately with clinicians' ratings of their clients' anxiety levels (Buss, Weiner, Durkee, & Baer, 1955). Palmer sweating (a physiological operational definition of anxiety) is greater for high-anxious subjects than low-anxious subjects in a verbal conditioning experiment (Haywood & Spielberger, 1966).

There is also evidence that the MAS correlates significantly with social desirability (Crowne, 1979; Hagborg, 1991). Subjects low on the MAS may desire to make a better impression on others than high MAS scorers. This makes us wonder to what degree the MAS is also reflective of social desirability, the desire to portray oneself in a positive light.

While the MAS measures one's general level of anxiety, you know that your anxiety level also fluctuates during the day. We may, in general, not feel anxious, but in certain circumstances or setting we may. Several researchers (e.g., Cattell & Scheier, 1961; Spielberger, 1966) responded to this observation by differentiating between two types of anxiety: trait and state anxiety. Trait anxiety is the relatively stable anxiety level of individuals, the construct that the MAS is supposed to measure. State anxiety is a situational anxiety level which fluctuates depending upon the situation in which you are. For instance, you have an enduring stable trait anxiety level. But, your state anxiety level while having a leisurely weekend breakfast would, in all likelihood, be much lower than just before you take an academic exam. The MAS is a confounded measure of trait and state anxiety. Spielberger (1977) has developed a State-Trait Anxiety Inventory. The form which assesses trait anxiety asks subjects to read 20 statements and indicate how they generally feel on a four point scale from "almost never" to "always." The form which assesses state anxiety asks subjects to indicate for 20 statements how they feel right now, at this moment, on a four point scale from "not at all" to "very much so." Below are examples of Trait and State Anxiety items from Spielberger's Inventory[1]:

TRAIT: Indicate how you GENERALLY feel
      I feel pleasant.
      I have disturbing thoughts.
      I am a steady person.

STATE: Indicate how you feel RIGHT now, that is, AT THIS MOMENT
      I feel calm.
      I am tense.
      I feel frightened.

Notice how the two scales differ. The Trait Anxiety scale is asking questions about how an individual feels most of the time. Whereas, the State Anxiety scale is asking how the individual is feeling right at the present.

Anxiety References:

Hagborg, W. J. (1991). The Revised Children's *Manifest Anxiety Scale* and social desirability. *Educational and Psychological Measurement, 51*, 423-427.

Bernstein, I. H., & Eveland, D. C. (1982). State vs. trait anxiety: A case study in confirmatory factor analysis. *Personality and Individual Differences, 3*, 361-372.

Buss, A. H., Weiner, M., Durkee, A., & Baer, M. (1955). The measurement of anxiety in clinical situations. *Journal of Consulting Psychology, 19*, 125-129.

Cattell, R. B., & Scheier, I. H. (1961). *The meaning and measurement of neuroticism and anxiety.* New York: Ronald Press.

Crowne, D. P. (1979). *The experimental study of personality.* Hillsdale, NJ: Lawrence Erlbaum Associates.

Harris, A. E., & Hanish, C. (1987). Anxiety and the span of apprehension. *Journal of Nervous and Mental Disease, 175*, 134-137.

Haywood, H. C., & Spielberger, C. D. (1966). Palmar sweating as a function of individual differences in manifest anxiety. *Journal of Personality and Social Psychology, 3*, 103-105.

Houston, B. K., Olson, M., & Botkin, A. (1972). Trait anxiety and beliefs regarding danger and threat to self esteem. *Journal of Consulting and Clinical Psychology, 38*, 152.

Ollendick, T. H., Yule, W., & Ollier, K. (1991). Fears in British children and their relationship to manifest anxiety and depression. *Journal of Child Psychology and Psychiatry and Allied Disciplines, 32*, 321-331.

Redding, C. A., & Livneh, H. (1986). Manifest anxiety: A cluster analytic study. *Perceptual and Motor Skills, 63*, 471-474.

Sarason, I. G. (1961). Characteristics of three measures of anxiety. *Journal of Clinical Psychology, 17*, 196-197.

Spielberger, C. D. (1966). Theory and research on anxiety. In C. D. Spielberger (Ed.), *Anxiety and Behavior*. New York: Academic Press.

Spielberger, C. D., Gorsuch, R. L., Lushene, P. R., Vagg, P. R., & Jacobs, G. A. (1968/1977). *State-Trait Anxiety Inventory*. Palo Alto, CA: Consulting Psychologists Press.

Taylor, J. A. (1950). The relationship of anxiety to the conditioned eyelid response. *Journal of Experimental Psychology, 41*, 81-92.

Taylor, J. A. (1953). A personality scale of manifest anxiety. *Journal of Abnormal Psychology, 48*, 285-290.

LABORATORY EXERCISE 4: PRACTICE IN DEFINING CONCEPTS OPERATIONALLY

This laboratory exercise is designed to give you practice in developing conceptual and operational definitions. For each of the following concepts, define them (a) conceptually and then (b) operationally.

1. aggression

   Conceptual:

   Operational:

2. obesity

   Conceptual:

   Operational:

3. dreaming state

   Conceptual:

   Operational:

4. memory

    Conceptual:

    Operational:

5. intelligence

    Conceptual:

    Operational:

6. attention

    Conceptual:

    Operational:

7. dominance

    Conceptual:

    Operational:

TOPIC 4: NATURALISTIC OBSERVATION

The laboratory goal for Topic 4 is to provide you with experience in (1) analyzing naturalistic observations, and (2) carrying out and evaluating your own naturalistic observations. You will be exposed to naturalistic observations with and without intervention (manipulation) from the experimenter.

LABORATORY EXERCISE 1: ANALYZING A NATURALISTIC OBSERVATION

Naturalistic observation involves recording subjects' naturally occurring behavior while they are in their natural environment. The subjects may be either humans or animals. The recording of natural behavior can be a very fruitful and worthwhile task which gives us much information about the distribution of phenomena in nature. For example, the well-known child psychologist Jean Piaget observed his children in all of their activities, writing down in great detail what he observed. From these observations he developed his theories about the development of cognitive and emotional stages in childhood. Another example is the work of Jane Gooddall in Africa. Over the years she has provided us with much new information about primates through her meticulous observations of these organisms in the wilds of Africa.

In this laboratory exercise you are to read an observation written by Wolfgang Köhler from his book, *The Mentality of Apes*. After reading the observation , you are to answer a series of questions which will help you understand how scientists can make use of behavioral data. The following description deals with events which were recorded over two years. To convey an accurate and in-depth account of their varied behaviors one must describe behavior covering a long period of time.

The everyday handling and treatment of objects on the part of the chimpanzee comes almost entirely under the rubric "play." If under the pressure of "necessity," in the special circumstances of an experimental test, some special method, say, of the use of tools, has been evolved -- one can confidently expect to find this new knowledge shortly utilized in "play," where it can bring not the slightest immediate gain, but only an increased "joie de vivre." And, on the other hand, one or other of the manipulations undertaken in the course of play can become of great practical utility. We will begin with a form of play that possesses this quality of utility (greatly overrated in Europe) in a high degree.

Jumping with the aid of a pole or stick was invented by Sultan, and first imitated, probably, by Rana. The animals place a stick, a long pole or a board upright or at a slight angle on the ground, clamber up it as quickly as possible with feet and hands, and then either fall with it in some direction, or swing themselves off from it in the very instant that it falls. Sometimes they spring to earth again, at other times onto a grating, beams, the branches of a tree, etc., often to a very considerable height. And at first it is certainly not circumstances that "forced the leap upon them." They could have "got there" much more easily by walking or climbing. Also, the landing-stages they selected seemed to offer no special attractions, so that when we take into account the constant repetition of this performance, we can only conclude that it is done out of the wish to jump and leap per se, just as children walk on stilts "for fun."

But very soon this form of play developed into the regular use of a tool. (Jan. 23, '14.) Sultan made the attempt to reach the objective (in the course of an experiment) in vain, as it was hung too high for him. He leapt straight into the air from the ground several times, and in vain; then he seized a pole that lay in his vicinity, lifted it as though to knock the prize down, and then desisting, pressed one end of the pole into the ground beneath the objective, and repeated the "climbing jump," as above described, several times in succession. His

37

movements had a certain playful and sketchy character, as though to say: It won't be any use!" and it was not. On the next occasion (February 3rd) he was more resolute and more fortunate; he approached a solid piece of plank, so heavy that he would only just cope with it, placed it in position and started climbing and jumping off. Three observers who were present maintained that he could not possible reach the prize in this manner, and on three occasions the treacherous plank fell over before he could swing himself away, but on the fourth trial he succeeded and tore down the fruit.

The use of the jumping-pole spread to Grande, Tercera and Chica and even to the heavy and clumsy Tschego, but skill and success with it varied greatly according to their individual ability. After some time Chica was easily first: she "jumped off" with the aid of short sticks and boards, and presently with a pole of over two metres long, which had appeared from somewhere. By its aid she could reach anything that was not more than three metres above the ground.

Later on, wishing to see how far her capabilities extended, I presented her with a bamboo over four metres long. She immediately showed completed mastery of this tool or toy, and climbed at frantic speed to a height of over four metres before the pole fell over. She herself at that time was not quite one metre tall, when drawn up to her full height. For certain reasons she had to be separated from her beloved toy for some time during the daytime; but in the evening, when she entered the playground where the bamboo lay, she repeatedly interrupted the (to her immensely important) business of a meal, in order to seize the coveted treasure and "just once" snatch a hasty jump.

[Of course this clever trick was only possible as a result of experience in placing the stick and controlling her own muscular efforts, in order not to lose balance before she had completed her climb. We must compare this to the achievements of a human gymnast: Chica has a "feeling for it." The draw-back is obviously the violent impact of a headlong fall from five metres onto hard piece of ground. Chica often inspects and touches those portions of her body which have borne the brunt of the fall and walks away with slow and subdued gait; but, thanks to her incomparable skills as a tumbler, she received no serious injuries. There was no "training" whatsoever about this, either: my part in the matter was solely the gift of the long bamboo. The jumping off was invented, introduced, further developed, and utilized to solve problems in the tests, by the chimpanzees themselves.

Imitation of human beings is excluded in this case. For, although acrobats may perform the same trick, there were none such in Tenerife and the ordinary pole-jumping of expert human acrobats is something quite different -- and not customary in the surroundings of the animals.][1]

Naturalistic Observation: Köhler's Chimpanzees

Answer the following questions on a separate sheet of paper and turn them into your instructor.

1. In what variables was Köhler interested?

2. List four objective behaviors which the chimpanzees performed and four interpretations Köhler made of the chimpanzees' behavior.

3. How might you do a similar study with children? What "play" behaviors would be interesting to observe in children between the ages of 2 and 4?

LABORATORY EXERCISE 2: DEVELOPING YOUR OBSERVATIONAL SKILLS

The purpose of this exercise is to help you develop your observational skills of a variety of interesting behaviors.

You are to walk around your campus for 15 minutes, both in the buildings and on the grounds and unobtrusively observe people's behaviors. That is, you are to observe people in such a manner that they do not know that they are being watched. The areas of the campus should be public so that the privacy of the individual is not violated.

As you observe people's behavior, develop a minimum of two experimental questions which, in your opinion, would be worthwhile for a behavioral scientist to investigate unobtrusively. Write them in the space provided below, along with a short justification of why each is a worthwhile question. (Examples: Do men and women carry their books differently? How many people walk down the center of the hall vs. walk along the wall?)

1. Experimental questions and justifications:

Optional: When you return to class, your instructor may divide the class into small groups. Within each group, each student is to present their experimental questions. As a group, choose one of the experimental questions and design an observational instrument for investigating this question. The group must decide how to operationally define the behavior under investigation. Each group is then to present the proposed experimental question, operational definition(s) of the behavior(s) under investigation, and the observation instrument(s) which will be used to the entire class.

LABORATORY EXERCISE 3: OBSERVING DYADIC CONVERSATION DISTANCES

The purpose of this laboratory exercise is to give you experience in naturalistic observation without intervention.

The acceptable distance for a conversation between adults is greatly affected by the cultural background of the participants and the gender of those involved (Hall, 1966; Patterson, Reidhead, Good, & Stopka, 1984; Sommer, 1969). The quality and type of interaction affects the distance we stand from one another in conversations. If we feel like we are good friends with another person, we will stand closer than if we feel like strangers. Observational studies in real life situations have found that individuals in Arab and Latin American countries, in general, stand significantly closer to one another during conversations than do Americans and Northern Europeans (e.g., Hall, 1966; Sanders, Hakky, & Brizzolara, 1985). Within the American culture, studies have shown that male-female dyads stand closest, female-female dyads are intermediate, and male-male dyads are most distant (e.g., Baxter, 1970; Sommer, 1962). By contrast, Remland, Jones, and Brinkman (1991), based upon video recordings of 253 naturally occurring dyadic interactions found differences between the Netherlands, France and England.

The present laboratory exercise was designed to examine the distances at which men and women interact with each other on your school campus. You are to observe dyads (pairs) of male-male, female-female, and male-female students conversing with one another. Determine in class where on your campus (e.g., the library, in the halls, outside, in the cafeteria, etc.) you will observe these dyads.

On the following page is provided an observational sheet. First, develop a hypothesis to investigate. Randomly choose dyads who are talking alone. Estimate and record the distances, in inches, between their faces (not other body parts). You should practice estimating distance prior to data collection.

After you have completed your observations, determine the mean and standard deviation for each of the three groups. You can conduct a t-test for independent groups to determine whether your groups are significantly different from one another. Since there are three groups involved, a more legitimate test would be a one-way analysis of variance with subsequent mean tests performed between the groups. You may or may not have the background to do the latter statistical test. Finally, after analyzing the data, interpret your results. Your instructor will direct you as to how to write up this study.

References:

Baxter, J. C. (1970). Interpersonal spacing in natural settings. *Sociometry, 33*, 444-456.

Fisher, J. D., & Bryne, D. (1975). Too close for comfort: Sex differences in response to invasions of personal space. *Journal of Personality and Social Psychology, 32*, 15-21.

Hall, E. T. (1966). *The Hidden Dimension.* Garden City: Doubleday.

Patterson, M. L., Reidhead, S. M., Gooch, M. V., & Stopka, S. J. (1984). A content-classified bibliography of research on the immediacy behaviors: 1965-82. Special Issue: Nonverbal intimacy and exchange. *Journal of Nonverbal Behavior, 8*, 360-393.

Sanders, J. L., Hakky, U. M., & Brizzolara, M. M. (1985). Personal space amongst Arabs and Americans. *International Journal of Psychology, 20*, 13-17.

Remland, M. S., Jones, T. S., & Brinkman, H. (1991). Proxemic and haptic behavior in three European countries. *Journal of Nonverbal Behavior, 16*, 215-232.

Sommer, R. (1962). The distance for comfortable conversation: A further study. *Sociometry, 25*, 111-116.

Sommer, R. (1969). *Personal Space.* Englewood Cliffs, NJ: Prentice-Hall.

Your Hypothesis:

Record the distance, in inches, between dyads in conversation. Choose only pairs who are talking alone with no others present.

Male-Male Dyad	Female-Female Dyad	Male-Female Dyad
1._____	1._____	1._____
2._____	2._____	2._____
3._____	3._____	3._____
4._____	4._____	4._____
5._____	5._____	5._____
6._____	6._____	6._____
7._____	7._____	7._____
8._____	8._____	8._____
9._____	9._____	9._____
10._____	10._____	10._____

When you have completed your observations, determine the following:

	Male-Male	Female-Female	Male-Female
Mean =	_____	_____	_____
Standard Deviation =	_____	_____	_____

Which dyad was the furthest apart?

> **intermediate?**

> **closest?**

Was your hypothesis verified?

You could conduct a statistical test on this data, according to your instructor's directions. Your instructor will direct you as to how to write up this study. Turn this observation sheet into your instructor.

LABORATORY EXERCISE 4: BOOK CARRYING BEHAVIOR

The purpose of this laboratory exercise is to give you experience in naturalistic observation without intervention. Observing the presence or absence of a behavior, or observing the presence of different types of behavior, is one of the simplest methods of recording behavior.

Everyday the majority of students carry books around campus, either in their arms or in a backpack. Interestingly, students usually adopt a characteristic mode of carrying behavior. On some campuses backpacks are very much the "in" mode -- the backpacks are carried in specific manners and may even be decorated with certain kinds of buttons or emblems.

Hanaway and Burghardt (1976) found gender differences in the book carrying modes of grade school, junior high, high school, and college students in the state of Tennessee. Jenni (1976; Jenni & Jenni, 1976) observed the book carrying behavior of men and women on 6 college campuses: 3 in the United States, 1 in Canada, and 2 in Central America (El Salvador and Costa Rica). In addition, they observed high school students and adults entering or leaving a public library in New York. Jenni observed all of the subjects unobtrusively, "never experimenting with or manipulating the subjects." She recorded "the carrying methods of all individuals who passed the defined point (a point chosen earlier by the experimenter) if they carried books without bags or packs" (p. 324).

Jenni classified the five most commonly observed carrying methods into two basic types (See Figure 1 below[2]): Type 1, in which the books partially clasped against the body and either one arm wraps around books (type A) or both arms wrap around books (type B); or Type 2, in which the books are held in one hand at the side of the body leaving the front of the body uncovered (types C, D or E).

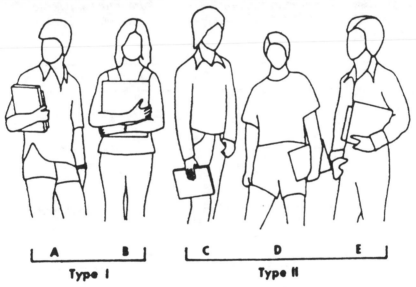

Using chi-square analyses, it was determined that there were significant gender differences in carrying behavior found in all colleges, both high schools, and the public library. Across the six college campuses, 82% of the women and 3% of the men were observed using Type I methods of bookcarrying, while 96% of the men and 16% of the women used the Type II methods. Only a small percentage used other methods which were categorized as Type III methods. Jenni (1976) noted that size or weight of books cannot adequately account for these gender differences, and suggested other possible contributing factors: social pressure and imitation, personality variables, sex differences in the relative accessibility of the body, and sex differences in the arrangement of the limbs. A limitation of observational research is that we cannot determine causes in the same manner that experimental research can often do.

Figure 1 (p. 325) reproduced with permission of author and publisher from: Jenni, M. A. (1976). Sex differences in carrying behavior. *Perceptual and Motor Skills, 43*, 323-330.

Your task is to design an unobtrusive observational study to replicate Jenni's research.

1. Determine where each experimenter will observe students (exs: entering or leaving classrooms, libraries, dormitories) and at what time (morning, afternoon, or evening).

2. Each experimenter is to designate a point at which when a student passes, a recording will be made. Record every student who is carrying books (but not a pack or bag) and who passes this point.

3. Prepare a recording sheet, or use the one provided below. Tally each subject under the appropriate column. If the bookcarrying behavior falls within another general category than Types I and II, check off Type III.

4. Once the data has been collected and collapsed across experimenters, you can do a chi-square analysis.

5. Your instructor will direct you as to how to write up this study.

References:

Alley, T. R., & Kolker, J. I. (1988). Psychological gender, hand preferences, and sex differences in book-carrying styles. *Perceptual and Motor Skills, 66*, 815-822.

Hannaway, T. P., & Burghardt, G. M. (1976). The development of sexually dimorphic book carrying behavior. *Bulletin of the Psychonomic Society, 7*, 267-270.

Jenni, M. A. (1976). Sex differences in carrying behavior. *Perceptual and Motor Skills, 43*, 323-330.

Jenni, D. A., & Jenni, M. A. (1976). Carrying behavior in humans: Analysis of sex differences. *Science, 194*, 859-860.

Book Carrying	Men	Women
Type 1: A:		
B:		
Type 2: C:		
D:		
E:		
Type 3: Others:		

LABORATORY EXERCISE 5: OBSERVING DRIVER BEHAVIOR

The purpose of this laboratory exercise is to give you experience in naturalistic observation with and without intervention. It has been hypothesized that drivers change their behaviors if they think they are being watched. You are to determine whether such a hypothesis is supported by your observations.

A natural place to observe drivers' behaviors is at a stop sign. The same stop sign should provide an observational environment that is relatively similar across a number of subjects, as long as there are no other environmental variables moderating the effect. Environmental variables which might moderate stop sign behavior are weather and traffic density.

You are to determine what drivers do when they are confronted with a stop sign under two conditions: (1) when they do not know that they are being watched, and (2) when they know that they are being watched.

You are to work alone, or with a partner. Complete the following steps:

Step 1: Operational Definitions

Operationally define what you mean by:

Full Stop:

Rolling stop:

No stop:

Step 2: Selection of Observation Location and Time Period

Select an area with a stop sign. This area must provide a location where you can unobtrusively, as well as obtrusively, observe drivers. Check with your classmates to determine that you will not be observing at the same time should you choose the same stop sign. Consider the time of the day as a variable to be controlled. You want to keep the traffic density approximately the same across your two conditions.

Step 3: How will you be obtrusive?

Determine how you will be unobtrusive and obtrusive. In other words, where will you stand? You may need to go to the stop sign area to determine if you have made the correct preparations.

Step 4: Collecting the Data

Collect the data by filling in the two recording sheets at the end of this exercise. One must be filled in when you are an unobtrusive observer and the other must be filled in while you are an obtrusive observer. If you are working in pairs, compare your observations.

On a separate sheet of paper, write down any general observations you make. Did it seem to make a difference whether there were other people in the car? Did anyone say anthing to you? If so, what?

Step 5: Summarizing Your Data

Now, from your observation sheets, summarize your data in the following table.

TYPE OF STOP	TOTAL SUBJECTS	MALE SUBJECTS	FEMALE SUBJECTS
UNOBTRUSIVE CONDITION: Full Stop N = Rolling Stop N = No Stop N =			
OBTRUSIVE CONDITION: Full Stop N = Rolling Stop N = No Stop N =			

Step 6: Analyzing Your Data

There are two directions you can take with your data: (1) descriptive statistics, and/or (2) inferential statistics. Your instructor will guide you as to what to do.

(1) **Descriptive statistics.** You can compute the percentage of individuals in each of the categories for the total sample, and for men and women separately. These percentages could also be compared between the obtrusive and unobtrusive conditions.

(2) **Inferential statistics.** You could apply the Chi Square analysis to determine if there is a significant difference between the two conditions as to whether they came to a full stop or not. Another analysis could compare men vs. women in each condition, or across conditions. Dependent upon what you want to know, you could compare the different types of stops in various ways.

Step 7: Interpreting Your Findings

Now that you have analyzed the data, you must interpret its meaning. Look at the computed percentages and/or Chi Square analyses. Decide what the data tells you about the behavior of drivers at stop signs. Hand in your observation sheets, summary table, and analyses.

Your instructor will direct you as to how to write up this study.

NAME: _____   **LAB:** _____

Date: _____   Time of Day: _____   Weather Condition: _____

UNOBTRUSIVE CONDITION

Observation	Gender of Driver 1 - Male 2 - Female	Type of Stop 1 - Full Stop 2 - Rolling Stop 3 - No Stop	Number of Persons in Vehicle	
			Driver Alone	With Passengers
1				
2				
3				
4				
5				
6				
7				
8				
9				
10				
11				
12				
13				
14				
15				
16				
17				
18				
19				
20				

Comments:

NAME: _____ **LAB:** _____

Date: _____ **Time of Day:** _____ **Weather Condition:** _____

OBTRUSIVE CONDITION

Observation	Gender of Driver 1 - Male 2 - Female	Type of Stop 1 - Full Stop 2 - Rolling Stop 3 - No Stop	Number of Persons in Vehicle	
			Driver Alone	With Passengers
1				
2				
3				
4				
5				
6				
7				
8				
9				
10				
11				
12				
13				
14				
15				
16				
17				
18				
19				
20				

Comments:

TOPIC 5: ARCHIVAL RESEARCH

The laboratory goal for Topic 5 is to provide you with experience in carrying out archival research.

Behavioral scientists do not always work with subjects, either observing their behavior, interviewing them, or manipulating conditions and recording how they respond. Much important information can be gathered indirectly through the study of records. There are many types of manipulations which cannot be imposed on individuals due to their ethical and/or procedural constraints. We cannot (and would not want to) impose natural disasters, job change, or death on individuals. Yet, records can let us study, retrospectively, a number of interesting behaviors.

Recorded information, both numerical and verbal, about individuals can be of interest to behavioral scientists. Statistical records are kept relating to many variables: (a) socioeconomic information, such as age, family size, and residence; (b) health statistics, such as birth and death rates; (c) public and private statistics, such as wages, hours of work, and financial transactions; and so on. Statistical records can be used as social indicators. Other archival sources are public or personal written documents: presidential talks, Congressional records, letters, diaries, school compositions, and so on. Another type of archival source is mass communications: magazines, newspapers, radio, and television. All of these sources can provide researchers with rich information that can be used to investigate a variety of research questions.

We highly recommend to you the following references on unobtrusive behavioral research:

Sechrest, L., & Belew, J. (1983). Nonreactive measures of social attitudes. *Applied Social Psychology Annual, 4*, 23-63.
Webb, E. J., Campbell, D. T., Schwartz, R. F., Sechrest, L., & Grove, J. B. (1981). *Nonreactive measures in the social sciences*, 2nd Edition. Boston, MA: Houghton Mifflin.

LABORATORY EXERCISE 1: VIOLENCE IN TELEVISION PROGRAMS

The purpose of this laboratory exercise is to provide you with experience in performing a content analysis of verbal and physical violence portrayed on television programs.

For some years behavioral scientists have been concerned whether watching violence on television can increase viewers' aggressive behavior. While controversial, laboratory studies and some naturalistic studies tend to confirm a relationship between watching violent television and aggressive behavior (e.g., Eron & Huesmann, 1987: Huesmann, 1982; NIMH, 1982). A U.S. Surgeon General's report on television and social behavior (NIMH, 1982) suggested that there are moderating factors (e.g., Wiegman, Kuttschreuter, & Baarde, 1992). After reviewing research supporting and not supporting a link between viewing violence on television and in movies, Lande (1993) proposed there is a small group of vulnerable viewers.

The majority of TV content analysis research dealing with the prevalence of violence has concentrated upon physical violence (for a review, see Signorielli, Gross, & Morgan, 1982). Rates of five or six violent acts per hour on prime time TV has been found in most countries (for review, see Cumberbatch, Jones, & Lee, 1988).   Yet, violence may also be verbal in content. In a study of the portrayal of aggression on North American (U.S. and Canada) television, Williams, Zabrack, and Joy (1982) reported that an average of nine acts of physical aggression and 7.8 acts of verbal aggression per program hour were observed. Certain modes of aggression were more common than others. Physical aggression involving the body or a weapon accounted for 48.5% of all aggressive behavior observed. The proportion of acts using the other modes were as follows: physical threat, 9%; verbal threat, 18.25%; verbal abuse, 15%; sarcasm, 8.9%; political/socioeconomic, 0.9%; passive aggression and harassment, 1%; and symbolic/joking, 7.4%. In most instances, aggression was portrayed as a successful means to resolve conflict.

Williams et al. (1982), among other things, examined the overall violence mean ratings for various types of programs. They had coders rate the degree of violence (defined as "physical or psychological injury, hurt, or death" (p. 369)) on a scale from 1 (not at all violent) to 7 (very violent). The mean ratings for the various types of programs were the following: adventure programs, 3.6; animated programs (cartoons), 5.2; documentary programs, 4.6; drama and medical programs, 1.9; game programs, 1.0; instruction and religion programs, 1.8; music, variety, and talk programs, 2.2; and situation comedy programs, 2.8. These researchers concluded that there is much evidence for a link between television and viewers' beliefs about social reality. Their coders "agreed that 25% of the programs carried the message 'The world is a dangerous place to be.'" (p. 377). They use schema theory to help explain the impact television's content has on viewers' beliefs. As they note,

> . . . A schema is a cognitive model (belief or concept) that is built up and modified as information is acquired. In general, its complexity will vary directly with the amount of relevant information and experience the individual acquires. Once a schema is established, it is used to process new information in what amounts to a matching process; the individual searches cognitively for a schema into which to fit information she or he encounters. . . . Schema theory would help to explain why adults' beliefs about social reality are most open to influence by television when they are not actively involved in viewing (Hawkins & Pingree, 1982). In the low involvement situation, a crude fit of the schema to incoming information will be acceptable, whereas high involvement will result in a more careful attempt to match, and viewers will be more likely to notice discrepancies between their real-world experience and television's model of social reality, thereby rejecting television's model. . . . Television's impact . . . is strongest when other sources of relevant information are lacking. It also would explain why the evidence concerning television's influence on beliefs about social reality is strongest for areas of social reality related to violence; violence abounds on TV, is presented with little variation, and most viewers have restricted experience with it (pp. 377-378).

Williams et al. (1982) did their content analyses in 1977. Since then there has continued to be pressure from certain lay groups, and even United States Attorney General Janet Reno in 1993, upon television producers to reduce the amount of violence portrayed in their programs. Has it worked? You may wish to read one case study of how social science research has been used to improve children's television (Stipp, Hill-Scott, & Dorr, 1987). In this laboratory exercise, while limited in the number of observations you will be able to make, you and your classmates will have an opportunity to evaluate a selection of television programs. It is highly recommended that you read the study by Williams et al. (1982) before carrying out your own content analyses. We based the recommended approach in this exercise upon their study.

As a class, decide which types of programs you will evaluate: music videos, adventure programs, children's animated programs, children's nonanimated programs, crime programs, documentary programs, drama and medical programs, or situation comedy programs. It is recommended that you choose a maximum of three program types. For instance, you might compare children's animated and nonanimated programs, or different types of music videos on Music Television (MTV). After you have determined the type of programs you will evaluate, choose representative productions for each.

Determine whether you will watch one or more examples of each particular program. If you have two classmates analyze the same program, then you can examine inter-rater reliability. Coders are usually trained extensively before carrying out the actual research so that they are highly reliable; we realize that this is not possible in the present laboratory exercise.

Use the following definition of physical violence, which Williams et al. used: "'The overt expression of physical force (with or without weapon) against self or other, compelling action against one's will on pain of being hurt or killed, or actually hurting or killing. Must be plausible and credible; no idle threats, verbal abuse, or comic gestures with no credible violent consequences. May be intentional or accidental; violent accidents, catastrophes, acts of nature are included' (Gerber, Note 1)" (p. 366). Aggression may also be verbal threats or abuse, passive aggression and harassment, sarcasm, or joking.

On the following page, tally the number of times you see each mode of aggression on the program. Also rate the overall violence of each program on a 7- point scale from 1 (not at all violent) to 7 (very violent). With your classmates, collapse your data within each program category. Determine the mean rate of acts per program hour for each of the modes of aggression. Determine the overall mean violence rating for each category. Your instructor will direct you as to how to write up this study.

Television Violence References:

Cumberbatch, G., Jones, I., & Lee, M. (1988). Measuring violence on television. Special Issue: Violence on television. *Current Psychology Research and Reviews, 7* , 10-25.

Eron, L. D., & Huesmann, L. R. (1987). Television as a source of maltreatment of children. *School Psychology Review, 16*, 195-202.

Gerber, G. (1974). *Cultural indicators project: TV message analysis recording instrument (Rev. ed)*. Philadelphia, PA: Annenberg School of Communications, University of Pennsylvania.

Hawkins, R. P., & Pingree, S. (1982). Some processes in the cultivation effect. *Communications Research,7*, 193-226.

Hodge, B., & Tripp, D. (1986). *Children and television: A semiotic approach*. Stanford, CA: Stanford University Press.

Huesmann, R. (1982). Television violence and aggressive behavior. In *Television and behavior: Ten years of scientific progress and implications for the eighties (Vol. 2)*. Rockville, MD: NIMH.

Huston, A. C., Wright, J. C., Rice, M. L., Kerkman, D., & St. Peters, M. (1990). Development of television viewing patterns in early childhood: A longitudinal investigation. *Developmental Psychology, 26*, 409-420.

Lande, R. G. (1993). The video violence debate. *Hospital and Community Psychiatry, 44*, 347-351.

National Institute of Mental Health (NIMH). (1982). *Television and behavior: Ten years of scientific progress and implications for the eighties*. Rockville, MD: NIMH.

Sherman, B. L., & Dominick, J. R. (1980). Violence and sex in music videos: TV and rock 'N' roll. *Journal of Communication, 36*, 79-93.

Signorelli, N., Gross, L., & Morgan, M. (1982). Violence in television programs: Ten years later. In *Television and behavior: Ten years of scientific progress and implications for the eighties (Vol. 2)*. Rockville, MD: NIMH.

Stipp, H., Hill-Scott, K., & Dorr, A. (1987). Using social science to improve children's television: An NBC case study. *Journal of Broadcasting and Electronic Media, 31*, 461-473.

White, G. F., Garland, F., Katz, J., & Scarborough, K. E. (1992). The impact of professional football games upon violent assaults on women. *Violence and Victim, 7*, 157-171.

Wiegman, D., Kuttschreuter, M., & Baarda, B. (1992). A longitudinal study of the effects of television viewing on aggressive and prosocial behaviours. *British Journal of Social Psychology, 31*, 147-164.

Williams, T. M., Zabrack, M. L., & Joy, L. A. (1982). The portrayal of aggression on North American television. *Journal of Applied Social Psychology, 12*, 360-380.

Tally Sheet: Modes of Aggression on Television

	Program #1	Program #2	Program #3
Program			
Time/Date			
Length			
Number of Acts Involving: 1. Violence to Body			
2. Use of Weapon			
3. Physical Threat			
4. Verbal Threat			
5. Sarcasm			
6. Passive Aggression, Harassment			
7. Symbolic/Joking			
Total Number of Aggressive Acts			
Overall Violence Rate: 1 (none) to 7 (very violent)			

LABORATORY EXERCISE 2: GRAFFITI

The purpose of this laboratory exercise is to provide you with experience in performing a content analysis of graffiti.

Graffiti are the anonymous inscriptions, markings and pictures on surfaces produced privately but often for later public observation and response by others. While many view graffiti as a nuisance and the result of vandalism or aberrant behavior, others see it as a way to communicate and pass on information. Psychologists and sociologists have noted that graffiti serve a variety of purposes: "to teach, to pass on information, to voice discontent and rebel, and to assert one's individuality" (Roscoe & Evans, 1986, p. 221). Its value as an unobtrusive measure of human thought is valued in social science (Klofas & Cutshall, 1985; Webb, Campbell, Schwartz, Sechrest, & Grove, 1981).

Anderson and Verplanck (1983) categorized western world graffiti into three primary categories: tourist graffiti, inner-city graffiti, and restroom graffiti. In old archeological ruins, such as Egyptian pyramids and Mayan buildings, as well as in contemporary buildings and on natural landmarks such as rocks and trees, one finds tourist graffiti: names, dates, expressions of love, and even statements. American inner-city graffiti take three major forms: (1) "the special language of the ghetto youth, which is largely concerned with names and identify, and is characterized by constructions such as the reflexive 'as', in 'Clarence as Lefty' or 'Leftly as Clarence'"; (2) gang graffiti, "which serve both to define one's 'turf' and to ward off interlopers"; and (3) the "graffiti 'masters', the 'kings of the walls,' whose names or logos, often in colored splendor, adorn buildings, subway cars stations, bridges, trucks -- even elephants and airplanes" (Anderson & Verplanck, 1983, p. 342). In recent years there has been an increase of "Hip Hop" graffiti ranging from signature tags to elaborate polychrome spray-painted murals that can be studied to learn about the writers' values (Brewer & Miller, 1990). The most common graffiti are found in public restrooms.

Graffiti in public restrooms may be quickly scrubbed away by the janitors or left; some bar owners even provide blackboards to encourage patrons to write only there and not vandalize the walls. In general, more graffiti are found in male than female restrooms. Female graffiti make fewer sexual references and are usually more socially acceptable in regard to language and content (Arluke, Kutakoff, & Levin, 1987; for review, see Anderson & Verplank, 1983). Bruner and Kelso (1980) found restroom graffiti written by women to be more interpersonal, interactive and romantic, while men's graffiti tended to be more egocentric and contain more competitive statements about love.

While some research (e.g., Anderson & Verplank, 1983) is atheoretical, much of the early research was driven by Freudian theory and interpreted graffiti in terms of unconscious impulses, primitive thoughts, and infantile sexuality. More recently, some researchers take the perspective that graffiti "reflect the power positions of men and women in the social structure" (Bruner & Kelso, 1980, p. 250). Coming from a cognitive developmental perspective, Lucca and Pacheco (1986) interpreted children's graffiti in Puerto Rican schools in terms of its relationship to "the child's immediate life experiences such as concerns with their self- identity, interpersonal relations, cultural understandings, sexuality, and religious and political beliefs" (p. 465). Graffiti can be indicators of a community's social attitudes, such as towards war or homosexuality (Stocker, Dutcher, Hargrove, & Cook, 1972), and as such can show attitude differences in the patrons in different buildings on campuses, or bars in different socioeconomic areas of a town.

Given this as a background, your laboratory assignment is to collect graffiti from restrooms, or other public places, on your campus and/or in your town. Decide which buildings and public places will be surveyed, and assign students to collect the graffiti.

Record all graffiti in each place selected. You may wish to determine if the graffito is a reply to a previously existing graffito. After the data have been collected, each graffito must be categorized by content or theme. You may wish to develop your own categories, based upon a review of the literature and pilot observations. In a study of restrooms at the University of Tennessee in 1980, Anderson and Verplanck (1983) employed the following categories (percentages found in their study are provided in parentheses): political statements or references to political candidates (18%); humorous statements that are not targeted (14%); simple replies to other graffito that differs in content (11%); references to sex (11%); references to race (8%); general insults that defame a group or individual (7%); references to music (6%); references to religion (6%); references to fraternities or sororities (5%); references to drugs (3%); references to sports (2%); references to nuclear power or environmental issues (2%); statements that are philosophical in tone (2%); and miscellaneous which was less than 1% in a category (5%).

Rater reliability can be determined by having two students independently categorize the graffiti, or a subset of them, and then determine the percent of agreement between the raters. Anderson and Verplanck (1983) reported a rater reliability (percentage agreement) of 92%; while quite adequate it was not higher because some raters were unfamiliar with certain names, music or political issues.

For the purpose of comparison, Chi-square tests of significance can be used to determine differences between (1) male and female restrooms; (2) restrooms in different buildings on campus (for example, humanities and arts compared with engineering); (3) restrooms in buildings in different areas of a city; and/or (4) restrooms in public places which provide different services (for example, college campus and bars downtown). You might also wish to compare your findings with previous ones published in the literature (e.g., Anderson & Verplanck, 1983; Arluke et al., 1987; Bruner & Kelso, 1980) so that you can provide a historical perspective. Your instructor will direct you as to how to write up this laboratory exercise.

Graffiti references:

Anderson, S. J., & Verplanck, W. S. (1983). When walls speak, what do they say? *Psychological Record, 3*, 341-359.

Arluke, A., Kutakoff, L., & Levin, J. (1987). Are the times changing? An analysis of gender differences in sexual graffiti? *Sex Roles, 16*, 1-7.

Brewer, D. D., & Miller, M. L. (1990). Bombing and burning: The social organization and values of Hip Hop graffiti and implications for policy. *Deviant Behavior, 11*, 345-369.

Bruner, E. M., & Kelso, J. P. (1980). Gender differences in graffiti: A semiotic perspective. *Women's Studies International Quarterly, 3*, 239-252.

Greeson, L. E. (1990). Bus stop graffiti: An index of media based cultural instruction in a Scandinavian city. *Nordisk Psykologi, 42*, 358-369.

Klofas, J. M., & Cutshall, C. R. (1985). The social archeology of a juvenile facility: Unobtrusive methods in the study of institutional cultures. Special Issue: Innovative sources and uses of qualitative data. *Qualitative Sociology, 8*, 368-387.

Lucca, N., & Pacheco, A. M. (1986). Children's graffiti: Visual communication from a developmental perspective. *Journal of Genetic Psychology, 147*, 465-479.

Roscoe, B., & Evans, J. A. (1986). Desk top graffiti. *College Student Journal, 20*, 221-224.

Webb, E. J., Campbell, D. T., Schwartz, R. F., Sechrest, L., & Grove, J. B. (1981). *Nonreactive Measures in the Social Sciences*, 2nd Edition. Boston, MA: Houghton Mifflin.

TOPIC 6: SURVEYS

The laboratory goal for Topic 6 is to provide you with experience in evaluating published survey questionnaires, administering questionnaires, and analyzing and interpreting your questionnaire results. The first laboratory exercise requires you to evaluate published survey questionnaires. The subsequent exercises provide anonymous survey questionnaires which you may administer. Your instructor will assign the ones you are to complete.

It is essential that the responses to each survey remain completely anonymous. It is recommended that you stand away from your respondent while he or she fills out the questionnaire. Also have the respondent put the questionnaire into a manila envelope to insure anonymity.

LABORATORY EXERCISE 1: EVALUATING SURVEYS

The purpose of this laboratory exercise is to provide you with experience in evaluating published survey questionnaires.

Bring to class any questionnaire you can find. Some common sources include: (1) popular magazines, such as *Psychology Today, Time*, or *Newsweek*; (2) newspapers, either your local paper or a national paper such as *USA Today*; (3) a national census questionnaire that can be obtained in the U.S. Documents section of your library; or (4) a national poll done by a private pollster company, such as Gallop Poll.

As a class, or in small groups, evaluate your questionnaires as to possible biases in the questions and how well the questions are written. Were there any biases in the sampling procedures? If so, how might they influence the validity of the survey and the degree to which the results can be generalized to other populations? Were there any biases in the reporting of the results?

When you evaluate the questions themselves, ask the following:

1. Can the question be misunderstood? Is it too difficult for the less educated respondent?

2. Does the question assume too much knowledge upon the part of the respondent?

3. Is the question too long?

4. Does the question have double negatives?

5. Is the wording of the question biased or slanted towards a certain viewpoint?

6. Is the wording of the question leading? In other words, does it try to put words into the mouth of the respondent?

7. Is the best response format used? For instance, are enough alternatives provided if it is a fixed-alternative item. Or, would the question be better as a dichotomous response (example: true or false) or as a multiple-choice response (example: always, usually, sometimes, rarely, never)?

List below at least three biased and/or poorly written questions from the questionnaires you evaluated. Explain why each is biased or poorly written.

Rewrite the above questions so that they do not contain the flaws you identified.

LABORATORY EXERCISE 2: SURVEY ON ALCOHOL USE

The purpose of this laboratory exercise is to give you experience in conducting a survey. You may wish to present a summary of your findings to your counseling center, health center, or Dean of Students.

The use of alcohol and its related problems among college students is of major interest to researchers, counselors and administrators on almost any college campus. While alcohol use is quite prevalent in both colleges and high schools its prevalence is moderated by demographic variables (e.g., Skager & Fisher, 1989; Maddahian, Newcomb & Bentler, 1986). Recent studies (e.g., Barnes, Welte, & Dintcheff, 1992; Engs & Hanson, 1985; Haworth-Hoeppner, Globett, Stem, & Morasco,1989; Kraft, 1985; O'Hara, 1990) have found that similar percentages of college men and women report alcohol consumption, but that men continue to be more frequent and heavier users of alcohol. This may be due in part to women perceiving a greater risk with use of alcohol than men (Spigner, Hawkins, & Loren, 1993). Hilton (1987) reported that 12% of men aged 18-29 got drunk at least once a week. One in five living in a fraternity or sorority were heavy drinkers in a recent university survey (Haworth-Hoeppner et al., 1989). Alterman et al. (1990) reported that nearly half of their sample of college men experienced two or more drinking-related adverse consequences and over a third were intoxicated four or more times monthly within the past year.

Comparisons across studies can be confounded by differential indices for the categorization of alcohol drinking levels. Early researchers often depended upon frequency of alcohol use to define alcohol usage categories, while more recently often a combination of both frequency of use and the typical quantity of alcohol consumed at any one occasion is used. For instance, Haworth-Hoeppner et al. (1989) reported (p. 840) the following categories in their study of over 1,000 randomly chosen students at a moderately sized (15,000-16,000), state-supported Southern university:

Nondrinker:	those who drink less than once a year or not at all (total: 16%; men: 15%; women: 17%)
Infrequent:	those who drink at least once a year, but less than once a month (total: 10%; men: 7%; women: 12%)
Light:	those who drink at least once a month,but no more than 1-3 drinks per occasion (total: 14%; men: 11%; Women: 17%)
Moderate:	those who drink at least once a month with no more than 4-5 drinks, once or twice a week with 1-3 drinks per occasion, or 5 times per occasion, or 5 times a week with 1 or less drinks per occasion (total: 24%; men: 20%; women: 27%)
Moderate-Heavy:	those who drink at least once a month with 6+ drinks per occasion, once or twice a week with 4-5 drinks per occasion, 4 to 5 times a week with 2-3 drinks, or 4 times a week with 2-3 drinks, or 4 times a week with 4-5 drinks per occasion (total: 22%; men: 25%; women: 20%)
Heavy:	those who drink once or twice a week with 6+ drinks per occasion or 5 times a week with 4-5 drinks per occasion (total: 14%; men: 22%; women: 7%)

Complications accompanying alcohol usage increase substantially in moderate to heavy alcohol users (e.g., Alterman et al., 1990; Haworth-Hoeppner et al., 1989; Pang, Wells-Parker, & McMillen, 1989). In Table 1 are presented the percentages of college students who reported complications within the previous year in Haworth-Hoeppner et al.'s survey (1989).

TABLE 1

Percentage of College Drinkers who have Complications*

Complications	Percentage of Drinker Types				
	Infrequent	Light	Moderate	Moderate-Heavy	Heavy
Had hangover	23	44	80	91	93
Nauseated/vomited	16	28	45	61	68
Driven after several drinks	22	37	66	87	89
Drunk while driving	12	21	47	74	81
Driven after knowing had too much to drink	22	37	66	87	89
Arrested DWI	1	2	2	3	7
Gone to class after several drinks	2	7	10	25	39
Missed class because of hangover	3	16	38	67	69
Criticized for drinking	6	6	13	18	23
Thought might have drinking problem	4	3	3	10	27
Got lower grade	1	7	7	14	25
Got into fight	2	5	9	18	25
Damaged property	2	3	11	13	27
Trouble with law	1	3	3	9	21

* Adapted from Haworth-Hoeppner et al., (1989, p. 848-849, Table 3) with permission from authors and by courtesy of Marcel Dekker Inc.

In a survey of over 4,000 university students between 1986 and 1988, Gonzalez (1989) found that those subjects "who started to drink in elementary and middle school reported significantly higher levels of consumption and problems than Ss who started drinking while in high school or college" (p. 225).

In another study (Alterman et al., 1989) of college men at a large private university in northeastern United States, 40% were categorized as problem drinkers (intoxicated 2 or more times per month and having suffered at least two adverse consequences within the past year), yet very few thought they had a problem. "The most commonly reported adverse consequences were: blackouts- 53%; missed time from school or work due to drinking- 41%; binges of two or more days- 23%; and the shakes- 21%" (p. 99). Sweeney (1989) provides a review of data about the alcohol- memory disturbance of blackouts, and how their relate to our current thinking about memory processes.

Other important areas of research assess the correlates of alcohol abuse and attitudes towards alcohol consumption. There are thousands of studies in the literature on either of these general topics. Their findings guide us in the development of laws, alcohol prevention programs, and therapeutic approaches to treating alcoholism.

Your task is to conduct a survey of alcohol usage and alcohol-related problems on your campus. In Appendix A you will find five copies of a questionnaire we developed, based upon the research of Haworth-Hoeppner et al. (1989) and others. You may make additional copies.

It is extremely important that the survey be anonymous and that you have received approval from your human subjects committee. Develop a preliminary statement that explains the survey, states who is conducting it, states it is completely anonymous, and explains how to return the survey so that it is treated anonymously. Your teacher and human subjects committee will provide your school's guidelines. In Topic 14, Laboratory Exercise 2, is a sample consent form for an anonymous survey that was conducted by Crawford's research methods class.

A number of hypotheses based upon prior studies can be investigated with the survey's data. For instance, based upon the above review of several recent college studies, you can generate several hypotheses, such as

1. An equal number of men and women will report drinking alcohol in the past year.

2. Men will report substantially heavier drinking patterns than women.

3. Those individuals who report moderate to heavy drinking patterns will report alcohol-related problems.

4. Those individuals who report early drinking ages will report more moderate to heavy drinking patterns than will those who started drinking at a later age.

The chi-square is an appropriate statistic to use to assess the above hypotheses. Note that you will need to determine how you will operationally define your categories of alcohol use.

Your teacher will instruct you as to how to write up this survey study.

Alcohol References:

Alterman, A. I., Hall, J. G., Purteill, J. J., Serales, J. S., Holahan, J. M., & McLellan, A. T. (1990). Heavy drinking and its correlates in young men. *Addictive Behaviors, 15*, 95-103.

Barnes, G. M., Welte, J. W., & Dintcheff, B. (1992). Alcohol misuse among college students and other young adults: Findings from a general population study in New York State. *International Journal of the Addictions, 27*, 917-934.

Desiderato, L. & Crawford, H. J. (in press). Risky sexual behavior in college students: Relationships between number of sexual partners, disclosure of previous risky behavior, and alcohol use. *Journal of Youth and Adolescence*.

Engs, R. C., & Hanson, D. J. (1985). The drinking patterns and problems of college students. *Journal of Alcohol & Drug Education, 31*, 65-83.

Gonzalez, G. M. (1989). Early onset of drinking as a predictor of alcohol consumption and alcohol-related problems in college. *Journal of Drug Education, 19*, 225-230.

Goodwin, L. (1992). Alcohol and drug use in fraternities and sororities. *Journal of Alcohol and Drug Education, 37*, 52-63.

Hanson, D. J., & Engs, R. C. (1986). Correlates of drinking problems among collegians. *College Student Journal, 20*, 141-146.

Haworth-Hoeppner, S., Globetti, G., Stem, J., & Morasco, F. (1989). The quantity and frequency of drinking among undergraduates at a Southern university. *International Journal of the Addictions, 24*, 829-857.

Hilton, M. E. (1987). Drinking patterns and drinking problems in 1984: Results from a general population survey. *Alcoholism, 11*, 167-75.

Kaplan, M. S. (1979). Patterns of alcoholic beverage use among college students. *Journal of Alcohol & Drug Education, 24*, 26-40.

Maddahian, E., Newcomb, M. D.,& Bentler, P. M. (1986). Adolescents' substance use: Impact of ethnicity, income, and availability. In B. Stimmel (Ed.), *Alcohol and substance abuse in women and children* (pp. 63-78). New York: Hayworth Press.

O'Hare, T. M. (1990). Drinking in college: Consumption patterns, problems, sex differences. and legal drinking age. *Journal of Studies on Alcohol, 51*, 536-541.

Pang, M. G., Wells-Parker, E., & McMillan, D. L. (1989). Drinking reasons, drinking locations, and automobile accident involvement among collegians. *International Journal of the Addictions, 24*, 215-227.

Skager, R., & Fisher, D. G. (1989). Substance use among high school students in relation to school characteristics. *Addictive Behaviors, 14*, 129-138.

Spigner, C., Hawkins, W., & Loren, W. (1993). Gender differences in the perception of risk associated with alcohol and drug use among college students. *Women Health, 20*, 87-97.

Sweeney, D. F. (1989). Alcohol versus Mnemosyne: Blackouts. *Journal of Substance Abuse Treatment, 6*, 159-162.

LABORATORY EXERCISE 3: SURVEY ON EATING DISORDERS

The purpose of this laboratory exercise is to give you experience in conducting and analyzing a survey regarding eating disorders. You may wish to present a summary of your findings to your counseling center, health center, or Dean of Students.

Both the public sector and various health disciplines have been concerned with the growing number of people who are experiencing eating disorders, namely anorexia nervosa and bulimia. They have become increasingly prevalent during the last decade. This is partially attributed to societal pressures to control one's weight (Streigel-Moore, Silberstein, & Robin, 1986) and social contagion (Crandall, 1988). Sometimes athletes may experience pressure to maintain certain weights that may lead to eating disorders (e.g., Oppliger, Landry, Foster, and Lambrecht, 1993; Sundgot-Borgen, 1993). Prevalence estimates have ranged from 2% to over 18% in women (Thelen, McLaughlin-Mann, Pruitt, & Smith, 1987). While no longer rare, these eating disorders are more prevalent among normal-weight women than men. According to the American Psychiatric Association's (1987) *Diagnostic and Statistical Manual of Mental Disorders*, 3rd edition (DSM-III-R), the essential features of Anorexia Nervosa are

> . . . refusal to maintain body weight over a minimal normal weight for age and height; intense fear of gaining weight or becoming fat, even though underweight; a distorted body image; and amenorrhea (in females). . .
> . . . People with this disorder say they "feel fat," or that parts of their body are "fat," when they are obviously underweight or even emaciated. They are preoccupied with their body size and usually dissatisfied with some feature of their physical appearance.
> The weight loss is usually accomplished by a reduction in total food intake, often with extensive exercising. Frequently there is also self-induced vomiting or use of laxatives or diuretics. (In such cases Bulimia Nervosa may also be present.) (p. 65)

The essential features of bulimia, according to DSM-III-R, are

> . . . recurrent episodes of binge eating (rapid consumption of a large amount of food in a discrete period of time); a feeling of lack of control over eating behavior during the eating binges; self-induced vomiting, use of laxatives or diuretics, strict dieting or fasting, or vigorous exercise in order to prevent weight gain; and persistent overconcern with body shape and weight. In order to qualify for the diagnosis, the person must have had, on average, a minimum of two binge eating episodes a week for at least three months.
> Eating binges may be planned. The food consumed during a binge often has a high caloric content, a sweet taste, and a texture that facilitates rapid eating. The food is usually eaten as inconspicuously as possible, or secretly. The food is usually gobbled down quite rapidly, with little chewing. Once eating has begun, additional food may be sought to continue the binge. A binge is usually terminated by abdominal discomfort, sleep, social interruption, or induced vomiting. . . (p. 67)

Various self-report questionnaires have been devised to assess symptoms of these eating disorders. The Eating Attitudes Test (EAT; Garner & Garfinkel, 1979) and the Eating Disorder Inventory (Garner, Olmsted, & Polivy, 1983) assess the cognitive and behavioral characteristics of eating disorders. These questionnaires have been used to identify abnormal eating patterns among college students (e.g., Berg, 1988; Hart & Ollendick, 1985; Lundholm & Wolins, 1987). An excellent self-report questionnaire which assesses symptoms of bulimia is the BULIT (Smith & Thelen, 1984) or BULIT-R (Thelen, Farmer, Wonderlich, & Smith, 1991).

Your task is to conduct a survey of a sample of students on your campus to determine the distribution of bulimia among men and women. You will use the BULIT-R. Five copies of the BULIT-R can be found in Appendix A. Eight unscored items have been removed from this version. If you need more copies, you have permission to make photocopies. In class, determine how your class can best randomly sample your campus. Your instructor will direct you as to how to write up this survey study.

In addition to examining percentages of responses to each of the categories between men and women, you could do further analyses by weighing the responses as Thelen et al. (1990) did. All BULIT-R items are present in a 5-point, forced choice Likert format. Five points are given for the extreme "bulimic" direction, down to one point for the extreme "normal" direction. Some items are reversed in order to prevent a response bias. Those items for which $a=1$, $b=2$, $c=3$, $d=4$, and $e=5$ are: 1, 3, 4, 8, 16, 18, 20, 21, 26, and 27. Those items for which $a=5$, $b=4$, $c=3$, $d=2$, and $e=1$ are: 2, 5, 6, 7, 9, 10, 11, 12, 13, 14, 15, 17, 19, 22, 23, 24, 25 and 28. Thelen et al. (1990) employ a cutoff of 104 or above to meet the scale criteria for bulimia, although they suggest lowering it to a cutoff of 85 to reduce the number of false negatives.

Eating Disorders References:

American Psychiatric Association. (1987). *Diagnostic and Statistical Manual of Mental Disorders* (Revised 3rd ed). Washington, D.C.: American Psychiatric Association.

Berg, K. M. (1988). The prevalence of eating disorders in co-ed versus single-sex residence halls. *Journal of College Student Development, 29*, 125-131.

Brelsford, T. N., Hummel, R. M., & Barrios, B. A. (1992). The Bulemia Test - Revised: A psychometric investigation. *Psychological Assessment, 4*, 399-401.

Crandall, C. S. (1988). Social contagion of binge eating. *Journal of Personality and Social Psychology, 55*, 588-598.

Garner, D., & Garfinkel, P. E. (1979). The Eating Attitudes Test: An index of the symptoms of anorexia nervosa. *Psychological Medicine, 9,* 1-7.

Garner, E. M., Olmstead, M. P., & Polivy, J. (1983). Development and validation of a multidimensional eating disorder inventory for anorexia nervosa and bulimia. *International Journal of Eating Disorders., 2*, 15-34.

Hart, K. J., & Ollendick, T. H. (1985). Prevalence of bulimia in working and university women. *American Journal of Psychiatry, 142*, 851-854.

Howat, P. M., & Saxton, A. M. (1988). The incidence of bulimic behavior in a secondary and university school population. *Journal of Youth and Adolescence, 17* 221-231.

Lundholm, J. K., & Wolins, L. (1987). Disordered eating and weight control behaviors among males and female university students. *Addictive Behaviors, 12*, 275-279.

Oppliger, R. A., Landry, G. L., Foster, S. W., & Lambrecht, A. C. (1993). Bulimic behaviors among interscholastic wrestlers: A statewide survey. *Pediatrics, 91*, 826-831.

Smith, M. C., & Thelen, M. H. (1984). Development and validation of a test for bulimia. *Journal of Consulting and Clinical Psychology, 52*, 863-872.

Stein, D. M., & Brinza, S. R. (1989). Bulimia: Prevalence estimates in female junior high and high school students. *Journal of Clinical Child Psychology, 18*, 206-213.

Streigel-Moore, R. H., Silberstein, L. R., & Rodin, J. (1986). Toward an understanding of risk factors for bulimia. *American Psychologist, 41*, 246-263.

Sundgot-Borgen, J. (1993). Prevalence of eating disorders in elite female athletes. *International Journal of Sport Nutrition, 3*, 29-40.

Thelen, M. H., Farmer, J., Wonderlich, S., & Smith, M. (1991). A revision of the Bulimia Test: the BULIT-R. *Psychological Assessments: A Journal of Consulting and Clinical Psychology, 3* , 119-124.

Thelen, M. H., McLaughlin-Mann, L. M., Pruitt, J., & Smith, M. (1987). Bulimia: Prevalence and component factors in college women. *Journal of Psychosomatic Research, 31*, 73-78.

Welch, G., Thompson, L., & Hall, A. (1993). The BULIT-R: Its reliability and clinical validity as a screening tool for DSM-III-R bulimia nervosa in a female tertiary education population. *International Joural of Eating Disorders, 14*, 95-105.

LABORATORY EXERCISE 4: SURVEY ON ESP ATTITUDES AND EXPERIENCES

The purpose of this laboratory exercise is to give you experience in conducting a survey, analyzing its results, and comparing the results to those in the literature.

The existence of extrasensory perception (ESP) has been the center of great controversy in the field of science. In one survey (Palmer, 1979) over half of a sample of college students and townspeople in Charlottesville, Virginia, and surrounding communities reported that they had had at least one ESP experience. Many of the surveys about psychic experiences have been conducted on preselected samples or subsamples of the general population, such as college students. Others have attempted to obtain a more representative sample of the general population. Reports of having ESP experiences are, of course, not the same as demonstrating ESP under stringent laboratory controls. Yet, it is important to know how many people believe in ESP and how many people claim to have had ESP experiences, with a breakdown of various kinds.

Your task is to conduct a survey of attitudes towards ESP and of reported ESP experiences. In Appendix A there are 5 copies of a questionnaire on ESP which the authors developed for your use. You may photocopy more copies, if needed, for your study.

In class determine the group of people you will sample. Determine how you will randomly sample this group and how many questionnaires each of you will distribute.

Once you have conducted your survey, determine how you will analyze the data. Percentages of men, women, and total sample who answered yes to each statement can be determined easily. On the next page a table for your data summary is provided. Compare your findings with other surveys listed below. Write up the laboratory exercise according to your instructor's directions.

References:

Alvarado, C. S. (1987). Observations of luminous phenomena around the human body: A review. *Journal of the Society for Psychical Research, 54*, 38-60.

Blackmore, S. J. (1984). A postal survey of OBEs and other experiences. *Journal of the Society for Psychical Research, 52*, 225-244.

Haraldsson, E. (1985). Representative national surveys of psychic phenomena: Iceland, Great Britain, Sweden, USA and Gallup's multinational survey. *Journal of the Society for Psychical Research, 53*, 145-158.

Haraldsson, E., Gudmundsdottir, A., Ragnarsson, A., Loftsson, J., & Johnson, S. (1977). National survey of psychical experiences and attitudes toward the paranormal in Iceland. In J.D. Morris, W.G. Roll, R.L. Morris (Eds), *Research in Parapsychology 1976*. Metuchen, NJ: Scarecrow Press.

Johnson, R. D., & Jones, C. H. (1984). Attitudes towards the existence and scientific investigation of extrasensory perception. *Journal of Psychology, 117*, 19-22.

Palmer, J. (1979). A community mail survey of psychic experiences. *Journal of the American Society for Psychical Research, 73*, 221-251.

Rhine, L. E. (1956). Hallucinatory psi experiences. I. An introductory survey. *Journal of Parapsychology, 20*, 233-256.

Schmeidler, G. R. (1985). Belief and disbelief in psi. *Parapsychology Review, 16*, 1-4.

Table 1

Percentage of Respondents Claiming Belief in ESP or ESP Experiences

Statement		Men	Women	Total
BELIEFS				
in the existence of ESP	Yes			
	No			
ghosts exist	Yes			
	No			
life after death	Yes			
	No			
people can contact the dead	Yes			
	No			
flying saucers and people from other planets	Yes			
	No			
EXPERIENCES				
had an ESP experience	Yes			
	No			
telepathic experience	Yes			
	No			
precognitive dream	Yes			
	No			
out-of-body experience	Yes			
	No			
have seen a ghost	Yes			
	No			
experienced psychokinesis	Yes			
	No			
seen an aura	Yes			
	No			

TOPIC 7: EX POST FACTO STUDIES

The laboratory goal for Topic 7 is to provide you with experience in evaluating and conducting ex post facto studies. Ex post facto means "after the fact". In ex post facto studies the researcher is interested in the effects of previously determined traits, behaviors, or naturally occuring events on subsequent performance, attitudes, etc. The antecedent conditions can be genetically determined groups (e.g., gender, birth disorder), personality groups (e.g., introverts and extroverts), mental disorders (e.g., paranoid vs nonparanoid schizophrenics), life event conditions (e.g., divorced or raped), and so on.

The crucial point is that the antecedent condition cannot or should not be manipulated by the researcher. An ex post facto study looks like a true experiment but it is not since the researcher did not manipulate the antecedent conditions. Unlike a true experiment, there is no random assignment of subjects to conditions. The preexisting differences are the "manipulation", and the researcher is interested in the effect of these differences on subsequently measured variables. In addition, no conclusions of cause and effect can be made because some other variable other than that which was manipulated may have caused the relationship. Yet, like correlational studies, we can examine relationships between variables and learn much important information about correlates of preexisting individual differences.

LABORATORY EXERCISE 1: EVALUATING EX POST FACTO STUDIES

The purpose of this laboratory exercise is to provide you with experience in evaluating abstracts from published ex post facto studies. Your instructor may wish you to read one or more of these studies. Here your task is to identify (a) the groups, and subgroups, of subjects who were studied and on what criteria they were chosen, (b) the variables that were measured, (c) the results that were reported, and (d) any alternative factors, other than those studied, that may have contributed to the identified relationships. On a sheet of paper, write your answers for each selection and turn them into your instructor at the assigned time.

Barton, R. A., & Whiten, A. (1993). Feeding competition among female olive baboons, Papio anubis. *Animal Behaviour, 46,* 777-789.

Competition for food is thought to play a key role in the social organization of group-living primates, leading to the prediction that individual foraging success will be partly regulated by dominance relationships. Among adult females in a group of free-ranging olive baboons, dominance rank was significantly correlated with nutrient acquisition rates (feeding rates and daily intakes), but not with dietary diversity or quality, nor with activity budgets. The mean daily food intake of the three highest-ranking females was 30% greater than that of the three lowest-ranking females, providing an explanation for relationships between female rank and fertility found in a number of other studies of group-living primates. The intensity of feeding competition, as measured by supplant rates and spatial clustering of individuals, increased during the dry season, a period of low food availability, seemingly because foods eaten then were more clumped in distribution than those eaten in the wet season. Implications for models of female social structure and maximum group size are discussed.

Crawford, H. J., Brown, A. M., & Moon, C. (1993). Sustained attentional and disattentional abilities: Differences between low and highly hypnotizable persons. *Journal of Abnormal Psychology, 102,* 534-543.

Relations between sustained attentional and disattentional abilities and hypnotic susceptibility (Harvard Group Scale of Hypnotic Susceptibility: Form A; Stanford Hypnotic Susceptibility Scale: Form C) were examined in 38 low (0-3) and 39 highly (10-12) hypnotizable college students. Highs showed greater sustained attention on Necker cube and autokinetic movement tasks and self-reported greater absorption (Tellegen Absorption Scale) and extremely focused attentional (Differential Attentional Processes Inventory) styles. Hypnotizability was unrelated to dichotic selective attention (A. Karlin, 1979) and random number generation (C. Graham & F. J. Evans, 1977) tasks. Discriminant analysis correlated classified 74% of the lows and 69% of the highs. Results support H. J. Crawford and J. H. Gruzelier's (1992) neuropsychophysiological model of hypnosis that proposes that highly hypnotizable persons have a more efficient far frontolimbic sustained attentional and disattentional system.

Winokur, G., Coryell, W., Endicott, J., & Akiskal, H. (1993). Further distinctions between mani-depressive illness (bipolar disorder) and primary depressive disorder (unipolar depression). *American Journal of Psychiatry, 150,* 1176-1181.

Objective: Patients with bipolar disorder differ from patients with unipolar depression by having family histories of mania with an earlier onset and by having more episodes over a lifetime. This study was designed to determine whether additional aspects of course of illness, the presence of medical diseases, childhood traits, and other familial illnesses separate the two groups. Method: In a large collaborative study, consequently admitted bipolar and unipolar patients were systematically given clinical interviews. Data were collected on medical diseases and childhood behavioral traits. Systematic family history and family study data were also obtained. The patients were studied every 6 months for 5 years. Results: The group of bipolar patients had an earlier onset, a more acute onset, more total episodes, and more familial mania and were more likely to be male. These differences were relatively independent of each other.

Winocur, G. (1990). A comparison of cognitive function in community-dwelling and institutionalized old people of normal intelligence. *Canadian Journal of Psychology, 44,* 435-444.

Two carefully matched groups of normal old people living in institutions or in the community were administered a neuropsychological cognitive test battery. In general, the institutionalized group performed worse than the community group. Discriminant function analysis identified a subgroup of high-functioning institutionalized subjects whose performance more closely resembled that of the community group than the remainder of the institutionalized group. Differences between the various groups were not due to differences in IQ, age, health, or other controlled variables. The critical tests that differentiated the groups were sensitive to impaired function in frontal and medial-temporal lobe brain regions. The results suggest a complex interaction involving effects of age and environmental factors on brain function and cognition.

LABORATORY EXERCISE 2: GENDER DIFFERENCES IN SPATIAL ABILITY

The purpose of this laboratory exercise is to teach you how to do ex post facto research.

Before proceeding to read the material presented in this laboratory, your instructor will give you a timed visuospatial test that will be found in Appendix C. DO NOT LOOK AT THE TEST UNTIL YOUR INSTRUCTOR DIRECTS YOU!!!! As a class, read together the instructions and then do the sample problems. Your instructor will time you on the two parts of the test; each part takes 3 minutes. When you have completed the test, score it.

Male advantage for a variety of spatial ability tasks has been reported extensively (e.g., Maccoby & Jacklin, 1974; Vandenburg & Kuse, 1979), yet some meta-analyses (e.g., Caplan, MacPherson, & Tobin, 1985; Hyde, 1981) have suggested that such a conclusion is unwarranted because the proportion of variance accounted for by gender differences is less than 5%. When examined more closely, substantial and quite consistent gender differences occur on tasks requiring the manipulation of three-dimensional objects (e.g., Linn & Petersen, 1985; Vandenburg & Kuse, 1979). Evidence suggests that both environmental and genetic factors, in a complex interplay at different times and in different combinations, contribute to spatial ability development (Ashton & Borecki, 1987; Goldstein, Heldane, & Mitchel, 1990; Vandenberg & Kuse, 1979).

In this laboratory exercise, you have just taken the Mental Rotations Test (Vandenberg & Kuse, 1978). There is consistent evidence in the literature that men perform significantly better than women on this test. Your task is to determine whether there are significant gender differences on the Mental Rotations Test for your class. Your instructor will inform you as to how to do a t-test for independent groups and direct you as to how to write up this study. In addition, discuss how these differences might have come about. Do your parents have similar visuo-spatial abilities? As a child did you have a lot of experience manipulating things (e.g., puzzles, climbing trees, etc.)? What strategies did you use while you were doing the test (visualizing holistically vs verbalizing the parts)?

Spatial References:

Ashton, G. C., & Borecki, I. B. (1987). Further evidence for a gene influencing spatial ability. *Behavioral Genetics, 17,* 243-256.

Caplan, P.7 J., MacPherson, G. M., & Tobin, P. (1985). Do sex-related differences in spatial abilities exist? A multilevel critique with new data. *American Psychologist, 40,* 786-799.

Geary, D. C., Gilger, J. W., & Elliott-Miller, B. (1992). Gender differences in three-dimensional mental rotation: A replication. *Journal of Genetic Psychology, 153,* 115-117.

Goldstein, D., Haldane, D., & Mitchell, C. (1990). Sex differences in visual-spatial ability: The role of performance factors. *Memory and Cognition, 183,* 546-550.

Hyde, J. S. (1981). How large are cognitive gender differences? A meta-analysis using W^2 and D. *American Psychologist, 36,* 892-301.

Linn, M.A., & Petersen, A. C. (1985). Emergence and characterization of sex differences in spatial ability: A meta-analysis. *Child Development, 56,*1479-1498.

Maccoby, E. M., & Jacklin, C. N. (1974). *The psychology of sex differences.* Stanford, CA: Stanford University Press.

Vandenberg, S., & Kuse, A. R. (1978). Mental rotations: A group test of three-dimensional spatial visualization. *Perceptual and Motor Skills, 47,* 599-604.

Vandenberg, S., Kuse, A. R. (1979). Spatial ability: A critical review of the sex-linked major gene hypothesis. In M. Wittig & A. C. Petersen (Eds.), *Sex-related differences in cognitive functioning,* (pp. 67-95). New York: Academic Press.

For the Mental Rotations Test (Vandenberg & Kuse, 1978), use the following scoring instructions to determine your score after taking the test in class. The following scoring key presents each item and the possible four response figures. The two that are correct in each line are circled. Circle those figures which you correctly identified and put an X through those figures which you incorrectly identified. Do not evaluate any figure that you did not answer. In other words, we are only interested in the number of correctly marked figures and the number of incorrectly marked figures.

Part 1:

1.	①	2	③	4
2.	①	2	3	④
3.	1	②	3	④
4.	1	②	③	4
5.	①	2	③	4
6.	①	2	3	④
7.	1	②	3	④
8.	1	②	③	4
9.	1	②	3	④
10.	①	2	3	④

Part 2:

11.	1	②	3	④
12.	1	②	3	④
13.	1	②	3	④
14.	①	2	3	④
15.	1	②	3	④
16.	1	②	③	4
17.	①	2	③	4
18.	①	2	3	④
19.	1	②	3	④
20.	1	②	③	4

Determine how many correct and incorrect figures you identified. Remember not to include items which you did not answer.

	Part 1:	Part 2:	Parts 1 and 2 together:
correct:	___	___	___
incorrect:	___	___	___

Often cognitive tests take off for guessing by subtracting a certain percentage of the number wrong from the number right. For each problem, there are four possible answers, but only two are correct. The probability is that random responses would have given you a 50% chance of being correct (2 out of 4). Therefore, to correct for guessing, you are to subtract 1/2 of the number wrong from the number right. Calculate the corrected total score for Parts 1 and 2, separately and together.

Part 1:	Part 2:	Parts 1 and 2 together:
___	___	___

The MRT has two parts that were designed to be equivalent. If you were to do a two-factor (gender x part) analysis of variance, you can evaluate if there is a practice effect (Part 1 < Part 2) or an interaction between gender and part.

TOPIC 8: INDEPENDENT AND DEPENDENT VARIABLES

The laboratory goals of Topic 8 are to provide you with experience in both identifying and developing independent and dependent variables, as well as experience in carrying out research in which the independent variable is manipulated across two levels.

Until now, our topics have been concerned with research whose goals are descriptive or correlational in nature. We now turn our attention to the second major category of behavioral research: experimental research. Here the experimenter applies procedures that manipulate variables in a controlled manner so that we can demonstrate a cause-and-effect relationship among the variables. This topic, and the subsequent ones in the laboratory manual, addresses issues surrounding experimental research: independent and dependent variables, experimental control, control techniques, and experimental design.

As you will recall from reading your text, the independent variable represents the variable(s) or the antecedent conditions that are manipulated or varied by the experimenter. The dependent variable represents the behavioral variable designed to measure the influence of the independent variable. As such, these variables are of primary interest to the investigator because they determine if the hypothesized relationship does, in fact, exist. The independent variable is the hypothesized cause in the cause-and-effect hypothesis; the dependent variable is the effect subsequent to the application of the independent variable.

The independent variable can be manipulated in various ways. Presenting the independent variable to one group (experimental group) and not to the other (control group) is one common way. For instance, an experimenter studies the effect of music vs. the absence of music on work production. Another approach is manipulating the level (degree) of the independent variable. For instance, an experimenter manipulates the dose levels (low, medium, and high) of a drug and observes the subsequent effect on the chosen dependent variable, such as driving behavior. A third method is varying the type of independent variable. For instance, an experimenter may study the effects of candy vs verbal reward on children's performance on a task.

LABORATORY EXERCISE 1: IDENTIFYING INDEPENDENT AND DEPENDENT VARIABLES

In this laboratory exercise you are presented with a series of abstracts from published studies. Your task is to identify the independent variables and dependent variables. Identify the number of levels for each independent variable. On a sheet of paper, write your answers for each selection and turn them into your instructor at the assigned time.

Blonstein, R., & Geiselman, R. E. (1990). Effects of witnessing conditions and expert witness testimony on credibility of an eyewitness. *American Journal of Forensic Psychology, 8*, 11-19

Examined the influence of good vs poor witnessing conditions (WCs) and supportive vs unsupportive expert testimony (ET) on opinions of eyewitness credibility (EC). 100 undergraduate Ss read a case description containing various combinations of eyewitness testimony, WCs, and ET and rated the EC in each. Ss believed eyewitnesses significantly more under good WCs than poor WCs and also believed eyewitnesses significantly more with supportive than with unsupportive ET. An interaction was found between the type of ET and the change in credibility scores made by Ss, indicating an increase in credibility with supportive ET, and a decrease in credibility with unsupportive ET.

Carlson, C. L., Pelham, W. E., Milich, R., & Dixon, J. (1992). Single and combined effects of methylphenidate and behavior therapy on the classroom performance of children with attention-deficit hyperactivity disorder. *Journal of Abnormal Child Psychology, 20*, 213-232.

24 boys (aged 6-12 yrs) with attention deficit-hyperactivity disorder participated in an 8-wk treatment program. Each subject received placebo and 2 doses of methylphenidate (MPH: 0.3 mg and 0.6 mg) crossed with 2 classroom settings: (1) a behavior modification (BM) classroom, including a token economy system, time-out, and daily home report card and (2) a classroom setting not using these procedures. Dependent variables included classroom observations of on-task and disruptive behavior, academic work completion and accuracy, and daily self-ratings of performance. Both MPH and BM alone significantly improved Ss' classroom behavior, but only MPH improved academic productivity and accuracy. Singly, behavior therapy and 0.3 mg MPH produced roughly equivalent improvements in classroom behavior. The combination of behavior therapy and 0.3 mg MPH resulted in maximal BM, which was nearly identical to that obtained with 0.6 mg MPH alone.

Christensen, L., & Redig, C. (1993). Effect of meal composition on mood. *Behavioral Neuroscience, 107*, 346-353.

The influence of simple carbohydrate, complex carbohydrate, protein, and delayed meal conditions on plasma tryptophan ratios and mood of normal Ss was investigated. In Exp 1, 27 women consumed 1 of 4 meal conditions, had blood samples drawn, and completed mood assessment measures before and at 5 times during the next 3 hrs. Exp. 2 replicated Exp 1 except that blood samples were not drawn and an additional simple carbohydrate condition was included. The carbohydrate meal conditions elevated blood glucose levels and plasma ratio of tryptophan to other large amino acids over that of the protein condition. A decline in feelings of fatigue and distress occurred 30 min postprandial regardless of condition and persisted for the rest of the study.

Jocabs, L. F. (1992). Memory for cache locations in Merriam's kangaroo rats. *Animal Behaviour, 43*, 585-593.

The ability of Merriam's kangaroo rats, *Dipodomys merriami*, to remember the location of food caches and to relocate caches in the absence of the odour of buried seeds was examined. Eight wild-caught kangaroo rats cached seeds in an experimental arena, and retrieved them 24 h later. Before retrieval, all odours associated with the cache sites were removed and seeds were replaced in only half of the cache sites. During retrieval, kangaroo rats were significantly more likely to search cache sites, with or without seeds, than non-cache sites. Non-cache sites were primarily investigated after all cache sites had been searched, indicating that search of non-cache sites did not denote an error in cache retrieval. These results suggest that kangaroo rats can remember the locations of food caches, and can relocate cache sites even when there is no odour of buried seeds. To estimate the advantage enjoyed by the forager with greater information, a second experiment compared an owner's success in retrieving its caches with the success of naive kangaroo rats searching for these same caches. Nine wild-caught kangaroo rats were allowed to search for caches that were distributed in the same spatial pattern as that created by one kangaroo rat from the first experiment. The naive subjects found significantly fewer caches than had the cache owner in the same length of time. This suggests that the use of spatial memory by a Merriam's kangaroo rat to relocate its food caches gives it a competitive advantage over other kangaroo rats that may be searching for its caches.

Lickliter, R. (1990). Premature visual experience facilitates visual responsiveness in bobwhite quail neonates. *Infant Behavior and Development, 13,* 487-496.

Exposing precocial avian neonates to premature (prenatal) visual experience appears to accelerate postnatal intersensory functioning. For example, bobwhite quail chicks who received patterned visual stimulation as embryos require auditory and visual cues to direct their filial behavior earlier in postnatal development than do normally reared chicks. The mechanism(s) for this alteration in perceptual organization has, however, not been investigated. The present study examined whether accelerated postnatal intersensory functioning is the result of reduced species-specific auditory responsiveness and/or enhanced postnatal visual responsiveness. Results revealed that bobwhite quail embryos exposed to unusually early visual stimulation do not show reduced auditory responsiveness in the period immediately following hatching but do exhibit an accelerated pattern of species-typical visual functioning. Specifically, chicks who experienced patterned light during the last 24 to 36 hours prior to hatching were able to use visual cues to direct their species-specific social preferences earlier in postnatal development than were control chicks. This finding suggests that one result of unusually early visual stimulation is that subsequent behavior is organized to include the earlier-than-normal sensory information.

Nakajima, S., & Potvin, J. (1992). Neural and behavioural effects of domoic acid, an amnesic shellfish toxin, in the rat. *Canadian Journal of Psychology, 46,* 569-581.

To examine the neurotoxic effects of domoic acid, an amnesic shellfish toxin, electroencephalographic and behavioural experiments were conducted on 38 rats. Injection of domoic acid (0.5 - 1.0 mg/kg intravenously, or 0.04 - 0.08 μg intraventricularly) caused seizure discharges in the hippocampus, tonic-clonic convulsions, and death within a few days. Convulsions and ensuing death were prevented by diazepam. Animals pretreated with diazepam (5 mg/kg, ip) tolerated intraventricular dose of domoic acid 0.4 μg, but showed a loss of pyramidal neurons mainly in the CA3, CA4, and a part of CA1 areas of the dorsal hippocampus. Learning of a radial maze task was severely impaired in naive rats after intraventricular injection of domoic acid (and diazepam, ip). In the animals previously trained on the maze task, domoic acid interfered with relearning of the same task. These effects appear similar to those of kainic acid and are analogous to the symptoms observed in humans who ingested mussels tainted with domoic acid.

Rapkin, D. A., Straubing, M., & Holroyd, J. C. (1991). Guided imagery, hypnosis and recovery from head and neck cancer surgery: An exploratory study. *International Journal of Clinical and Experimental Hypnosis, 39,* 215-226.

The value of a brief, preoperative hypnosis experience was explored with a sample of 36 head and neck cancer surgery patients. 15 patients volunteered for the experimental hypnosis intervention. 21 patients who received usual care (no hypnosis) were followed through their hospital charts and were used as a comparison group. Hypnotic intervention and usual care groups were comparable in terms of relevant demographic variables. Postoperative hospitalizations for the hypnotic intervention group were significantly shorter than for the usual care group. Within the hypnotic intervention group, hypnotizability was negatively correlated with surgical complications and there was a trend toward a negative correlation between hypnotizability and blood loss during surgery. Findings suggest that imagery-hypnosis may be prophylactic, benefitting patients by reducing the probability of postoperative complications and thereby keeping hospital stay within the expected range. Recommendations are presented for a controlled, randomized, clinical trial with a sufficiently large sample to provide the opportunity for statistical analysis with appropriate power.

Schweizer, E., Patterson, W., Rickels, K., & Rosenthal, M. (1993). Double-blind, placebo-controlled study of a once-a-day, sustained-release preparation of Alprazolam for the treatment of panic disorder. *American Journal of Psychiatry, 150,* 1210-1215.

Objectives: The goals of this study were to assess the antipanic efficacy of a new, sustained-release formulation of alprazolam and to assess the safety and tolerability of once-a-day administration of 1 - 10 mg of sustained-release alprazolam. Method: One hundred ninety-four patients with a diagnosis of agoraphobia with panic attacks or panic disorder with limited phobic avoidance underwent a 1-week placebo washout before being randomly assigned to groups receiving 8 weeks of double-blind treatment with either sustained-release alprazolam or placebo. Results: There was a significant treatment effect favoring sustained-release alprazolam (highest mean dose = 4.7 mg/day) across almost all measures of anxiety, panic, and phobic avoidance, despite a significantly higher dropout rate in patients receiving placebo. Eighty-five percent of the patients treated with sustained-release alprazolam, compared with 61% of the patients given placebo, reported complete blockade of panic attacks by the end of 6 weeks of treatment. Sedation was the most commonly reported adverse effect. Discontinuation of sustained-release alprazolam was associated with moderate but transient levels of distress in 48% of the patients; discontinuation of placebo led to distress in only 10% of the patients. Nonetheless, there was no difference in the proportion of patients who were able to remain off the study drug for at least 2 weeks. Conclusions: These results suggest that sustained-release alprazolam is highly effective in the acute treatment of panic disorder at doses comparable to those in the originally marketed compressed tablet of alprazolam. The medication was well tolerated but showed rebound effects during a rapid drug taper after 6 weeks of treatment.

Stokes, A. F., Belger, A., Banich, M. T., & Taylor, H. (1991). Effects of acute aspartame and acute alcohol ingestion upon the cognitive performance of pilots. *Aviation, Space, and Environmental Medicine, 62,* 648-653.

Tested 13 pilots in a double-blind design using a cognitive test battery relating to perceptual-motor abilities, spatial abilities, working memory, attentional performance, risk taking, processing flexibility, and planning or sequencing ability. Ss were tested over 5 sessions consisting of pretest and posttest controls and 3 randomly ordered treatment sessions, in which Ss received aspartame, placebo, and ethyl alcohol (as the positive control). No detectable performance decrements were associated with the aspartame condition, but decrements in psychomotor and spatial abilities were detected in the ethanol condition. Results do not support the concerns expressed in anecdotal evidence regarding the deleterious effects of aspartame upon cognitive performance.Wood, T. B., & Lewis, S. N. (1987). The influence of alcohol and loud music on analytic and holistic processing. *Perception and Psychophysics, 41,* 179-186.

Wood, T. B., & Lewis, S. N. (1987). The influence of alcohol and loud music on analytic and holistic processing. *Perception and Psychophysics, 41,* 179-186.

Studied the effects of alcohol consumption and loud music on analytic and holistic information processing in 52 male undergraduates. A restricted classification task, an embedded figures task, and a concept learning task with the choice of analytic single attribute learning or a holistic approach were begun at blood alcohol levels of about 0.048% and completed at a mean of 0.84%. Music as a distractor was present in some conditions. Loud music reduced the analytic response in subjects consuming alcohol, but loud music increased it in subjects drinking a placebo. Alcohol decreased speed and embedded figure response. Although it slowed concept learning, alcohol did not decrease analytic responding. Results indicate that alcohol slows the rate of information processing.

LABORATORY EXERCISE 2: DEVELOPING INDEPENDENT AND DEPENDENT VARIABLES

The purpose of this laboratory exercise is to provide you with experience in developing independent and dependent variables.

Since most research is an outgrowth of prior research, this exercise represents a continuation of the abbreviated literature review which you conducted in Topic 2. Therefore, refer back to the area on which you conducted your abbreviated literature reviews (Topic 2) and assume you are going to conduct a research study on that topic. Since you have already reviewed the literature on this area and have developed a research question, you are now ready to complete the following steps, which focus on the independent and dependent variables.

STEP 1: From your research question identified in Topic 2, identify two independent variables that need to be manipulated to arrive at an answer to that research question. Operationally define them so that they are concrete and measurable.

STEP 2: After you have operationally defined the independent variables, you must establish variation in them. You must decide how you want to establish this variation for the two independent variables. Then you must construct different operations of the independent variables to correspond to the different levels of variation you have established. Naturally, the number of levels and number of independent variables used will vary, depending on your research problem. For each independent variable, describe in detail the variations you propose on the next page.

Independent Variable A

First Level or Type:

Second Level or Type:

Third Level or Type:

Independent Variable B

First Level or Type:

Second Level or Type:

Third Level or Type:

STEP 3: Now that you have specified how you plan to construct and vary each independent variable, it is frequently helpful to reassess each independent variable to determine if your operational definition really represents the conceptual variable you had in mind. You are now to identify ways in which you could assure yourself that your translation of the conceptual independent variable was appropriate. Your instructor can provide valuable assistance if you are unsure on how to complete this step. As a guide, remember that you are to ask yourself the following question: "How can I be sure that my operational definition of the independent variable will generate the effect on the subject that I want it to generate?"

STEP 4: Once the steps above are completed you can have reasonable assurance that you have devised good independent variables. The next task is to identify one or more good dependent variables that will measure the independent variables.

STEP 5: Now operationally define, in great enough detail that an experimenter could follow your description and replicate your study, each of the dependent variables identified in Step 5.

Dependent Variable(s) for Independent Variable A

Dependent Variable(s) for Independent Variable B

LABORATORY EXERCISE 3: PAIRED-ASSOCIATE LEARNING OF LOW AND HIGH IMAGERY WORDS

The purpose of this laboratory exercise is to teach you how to manipulate two levels of an independent variable in a within subjects design.

Mental imagery, defined as instructions to form a mental image of some material or by ratings of the image-arousing capacity of verbal material, has been shown to be related positively to performance on a number of memory tasks (for review, see Paivio, 1983). Think for a moment of the words "elephant" and "perhaps". What kinds of images and thoughts come to you? If you are like most individuals, the word "elephant" arouses many more images and associations than does the word "perhaps". Systematic research has shown that individuals can recall and/or recognize significantly more learned words that are high, than low, in imagery content. This has been found in studies where subjects learned lists of words that were presented only once or several times until a criterion was reached, or where they learned lists of paired-associate words (example: elephant-hammer) and then were presented with one of the pair (example: elephant) and asked to recall the other member of the pair.

Allen Paivio and his colleagues at the Department of Psychology, University of Western Ontario, Canada, have developed norms for imagery and familiarity ratings for thousands of nouns. Groups of subjects were presented a number of nouns. The subjects were instructed to rate the words on scales from 1 to 7, as to how much imagery was associated with each word. (They also rated other dimensions of the words such as concreteness and meaningfulness.) In addition, these words have been evaluated as to how frequently they occur in our everyday reading material. As researchers, we can use these norms to choose low and high imagery words that are matched on meaningfulness, frequency in our reading, and length of words, as well as other dimensions which may be of interest to us.

Your task in this laboratory exercise is to carry out a paired-associate word experiment, using subjects who are not your classmates. You will determine how well subjects recall paired-associate words that are either low or high in their image- arousing capacities. Based upon prior research, it is hypothesized that your subjects will recall significantly more high than low imagery words.

The stimuli necessary for this experiment are provided in Appendix B. They were taken from Paivio's (1981) Imagery and Familiarity ratings for 2448 words: unpublished norms. They were used as stimuli in a study conducted by one of your laboratory manual authors (Crawford, Allen, & Kiefner, 1983). The pairs of low or high imagery words were additionally matched for meaningfulness, frequency in the English language, and length of words.

For your experiment you are provided with 15 pairs of words which are high in imagery content and 15 pairs of words which are low in imagery content. In addition, there are 10 filler pairs, 5 of which are to be placed at the beginning of the series and 5 of which are to be placed at the end. These filler pairs are provided so as to decrease recency and primacy effects upon recall. On the subsequent pages of the Appendix are the cards which have only one of the two pairs typed on them. They will be used to trigger memory recall of the paired-associate word.

Before you begin the experiment, do the following:

1. Very carefully cut out all of the cards from Appendix B. Note that on the back of the cards there

is an indication as to whether they are filler cards, low imagery pairs, high imagery pairs, or recall stimuli.

2. Take the low and high imagery paired-associate cards and randomize them so that no more than three low or three high paired-associate cards are next to one another.

3. There are 10 filler cards. Put five of them as the first five of the stack and five at the end as the last 5 of the stack.

4. Now take the recall stimuli (one word on each card) and randomize them so that no more than three low or three high stimulus words are next to one another. Make a record of their order for later scoring of subject responses.

5. Take the response sheet which is at the end of this exercise. Make enough photocopies for your assigned subjects.

Now that you have your experimental materials ready, you can turn your attention to the procedure. In class decide how you will obtain your subjects and how many subjects you will test. Determine with your instructor whether you will need a consent form. It is helpful to practice the procedure with one of your classmates. You will need a watch to time the 10 second stimulus periods.

The recommended procedure is the following:

1. Tell your subject that this is an experiment concerning the learning of paired nouns (NEVER MENTION THAT YOU ARE LOOKING AT HIGH VS LOW IMAGERY WORDS) and that you are doing it for your experimental methods course. Add any other information your class decides on.

2. Sit down at a quiet table where there will be no interruptions. The subject is to sit across from the experimenter. Put a small barrier up so that the subject cannot read the information on the back of the cards.

3. Put the stimulus cards in one pile and the response cards in another pile.

4. Tell the subject the following: "In front of me I have a set of cards. On each card there are two nouns, such as "table" and "coffee". Your task is to learn that these two nouns are paired together so that when, at a later time, I give you another card that only has "table" on it you can recall the word "coffee" and say it to me. I will then write down your response. The procedure I will follow is this: First, I will give you a series of different cards, each for 10 seconds. On each card will be two words. You are to learn that these two words are paired together. After we complete this set of cards, I will turn to a second pile of cards in front of me. I will show you one card at a time for 10 seconds. During that time you are to tell me what word you think was associated with it. Always guess even if you cannot remember the associated word. Do not leave any blank. Are there any questions?" (If there are questions, go over the instructions again.)

5. Present the stimulus cards, one at a time for 10 seconds, in front of the subject in such a manner that the subject can easily read the card. Remove that card and present the next card. Continue until you have presented every card in the first pile (N=40).

6. Immediately turn to the second pile. Say to the subject: "Now I will present one of the two

pairs of words. Tell me within the 10 second period what the associated word is. If you are not sure, you must guess." Present the stimuli cards, one at a time for 10 seconds. Write down the subject's responses. (An alternate procedure is to have the subjects write down their own responses. A problem with this method is that the subjects have a longer, and more varied, length of time for writing their responses.) Note that the filler words are not in the second pile.

7. After the completion of the experiment, ask the subjects how they learned and recalled the words. Write down their comments. (Commonly used strategies are repeating the two words, making verbal associations with sentences or phrases, making imagery associations between the words, or nothing.)

8. Debrief the subject as to the purpose of the experiment.

Once you have completed testing your subjects, you are ready to analyze your data. For each subject, determine how many low imagery words and high imagery words were recalled correctly. You must only give them credit if the exact word is recalled. Do not give them credit for words that are only similar, no matter how close they are. Compute the means and standard deviations for the low and high imagery words recalled for all subjects. Your instructor may wish you to perform a t-test for non-independent groups so that you can determine whether the subjects recalled significantly more high than low imagery words. Your instructor will instruct you as to how to perform a t-test.

On a separate sheet of paper, present your results and interpret your findings, following the directions your instructor gives you. Be able to answer the following questions:

1. Is the hypothesis supported?

2. Did you find that some subjects recalled more words than others? If so, do you have any hypotheses as to why there were these individual differences?

3. How might you investigate these hypotheses in a future study?

Background References:

Crawford, H. J., Allen, S. A., & Kiefner, M. G. (1983). Effect of hypnosis on the retention of high- and low-imagery paired associates. *International Journal of Clinical and Experimental Hypnosis, 31,* 208.
Day, J. C., & Bellezza, F. S. (1983). The relation between visual imagery mediators and recall. *Memory and Cognition, 11,* 251-257.
Ernest, C. H., (1977). Imagery ability and cognition: A critical review. *Journal of Mental Imagery, 2,* 181-216.
Hampson, P. J., Marks, D. F., & Richardson, J. T. E. (Eds.) (1990). *Imagery: Current developments.* London: Routledge.
Marschark, M., & Hunt, R. R. (1989). A reexamination of the role of imagery in learning and memory. *Journal of Experimental Psychology: Learning, Memory, and Cognition, 15,* 710-720.
Paivio, A. (1965). Abstractness, imagery, and meaningfulness in paired-associate learning. *Journal of Verbal Learning and Verbal Behavior, 4,* 32-38.
Paivio, A. (1971/1983). *Imagery and verbal processes.* New York: Holt, Rinehart and Winston.
Richardson, J. T. E. (1980). *Mental imagery and human memory.* New York: St. Martin's Press.
Richardson, J. T. (1985). Converging operations and reported mediators in the investigation of mental imagery. *British Journal of Psychology, 76,* 205-214.
Straub, H. R., & Granass, M. M. (1992). Interaction of instructions with the recall strategy actually used in a paired-associates learning task. *Psychological Reports, 71,* 987-993.

Subject number _____   Gender: Male _____   Female _____

1.	16.
2.	17.
3.	18.
4.	19.
5.	20.
6.	21.
7.	22.
8.	23.
9.	24.
10.	25.
11.	26.
12.	27.
13.	28.
14.	29.
15.	30.

Subject's comments:

Number recalled correctly: Low imagery words _____   High imagery words _____

Response Recording Sheet: Paired-Associate Words

Subject number _____   Gender: Male _____   Female _____

1.	16.
2.	17.
3.	18.
4.	19.
5.	20.
6.	21.
7.	22.
8.	23.
9.	24.
10.	25.
11.	26.
12.	27.
13.	28.
14.	29.
15.	30.

Subject's comments:

Number recalled correctly: Low imagery words _____   High imagery words _____

LABORATORY EXERCISE 4: JURY SIMULATION RESEARCH

The purpose of this laboratory exercise is to teach you how to manipulate two independent variables, each of which have two levels of variation, in a factorial between subjects design.

The importance of the personal characteristics and socio-economic status of a defendant on the assessment of guilt by a jury member cannot be underestimated. Social psychologists and forensic psychologists have become interested in studying the effects of various factors on jury decision making. They have investigated the influence of personal characteristics of the defendant and/or the victim have on a jury's decisions of guilt or innocence. The influence of the characteristics of the jury members, individually and as a group, on the jury's decision have also been investigated. These research questions are addressed by an area of research referred to as jury simulation research.

Rather than study the actual jury process, although that is also studied by social psychologists, jury simulation research simulates courtroom situations through the use of questionnaires or asking experimental subjects to serve as simulated jury members while they watch videotapes of real or simulated trials. A major criticism of such research is that it is artificial. Asking experimental subjects, often introductory psychology students, to participate in jury decision making is not analogous to the real situation. In response to such criticism, some researchers have used as their subjects, individuals who have been called up to serve on a jury or who have recently served on a jury.

In this laboratory study, you will replicate a study carried out by David Landy and Elliot Aronson in 1969. It has been replicated and extended several times in the literature.

Two major references for you to read are the following, which can be found in your library.

Landy, D., & Aronson, E. (1969). The influence of the character of the criminal. *Journal of Experimental Social Psychology, 5,* 141-152.
Nemeth, C., & Sosis, R. H. (1973). A simulated jury study: Characteristics of the defendant and the jurors. *The Journal of Social Psychology, 90,* 221-229.

You can use *Citation Index* to find more recent articles which refer to the Landy and Aronson (1969) and Nemeth and Sosis (1973) studies.

Landy and Aronson (1969) studied college students' assessments of how many years a defendant should be sentenced to prison. The defendant, John Sander, was found guilty of having hit and killed a pedestrian while driving under the influence of alcohol. The victim is presented as either begin attractive or unattractive: a noted architect who is prominent in the community or a notorious gangster. The defendant is presented as either being attractive or unattractive. (In the second experiment there was also a neutral defendant whom we will not consider.) Subjects were in one of the following conditions:

(1) attractive victim; attractive defendant
(2) attractive victim; unattractive defendant
(3) unattractive victim; attractive defendant
(4) unattractive victim; unattractive defendant

Before reading the next paragraph, stop for a moment and consider the following questions. Under which experimental condition will college students give the longest prison term to the defendant? Under which experimental condition will they give him the shortest prison term?

After subjects read the description of the crime and the personal characteristics of the defendant and victim, they were "requested to consider the crime of negligent automobile homicide as punishable for from 1 to 25 years imprisonment, and to sentence the defendant to a specific number of years of imprisonment, according to their own personal judgement (p. 148)." After that they were asked several questions about how they felt about the defendant and the victim, using a nine-point scale from "extremely favorable" to "extremely unfavorable."

The means and standard deviations of the sentences, expressed as years of imprisonment, in the experimental conditions under consideration here are given below.

Defendant		Victim Attractive	Unattractive	Total
Attractive	N =	18	18	36
	M =	8.72	8.44	8.58
	SD =	4.18	6.60	5.45
Unattractive	N =	18	18	36
	M =	13.89	9.61	11.75
	SD =	5.76	5.98	6.18
Total	N =	36	36	
	M =	11.31	9.03	
	SD =	5.61	6.24	

In this factorial design we can look at the main effects for the two factors: (1) attractive vs. unattractive victim, and (2) attractive vs. unattractive defendant. We can also look at the interaction between the two factors.

Landy and Aronson (1969) reported that subjects in the unattractive defendant condition were more severe in their prison sentences than subjects in the attractive defendant condition (respectively, means of 11.75 and 8.58 years). The defendant who had an attractive victim received more prison years than the defendant with an unattractive victim. The unattractable defendant with an attractive victim received more prison years (13.89) than the three other conditions which did not differ significantly from one another.

In summary, this study demonstrates that the character of the defendant and the victim are important variables in how severely college students say they will sentence the defendant.

For the present study, you are to replicate this study. In Appendix B are copies of each of the four instructional conditions[1]. You may make additional copies. You are to ask males and females to answer these anonymous questionnaires about jury decision making. To keep it anonymous, have the subjects put the questionnaires into an envelope and send them to you. As a class, discuss where you will obtain your subjects (another class, people on campus randomly chosen, certain aged subjects, etc.).

---

[1]From Landy and Aronson (1989) with permission from author and Academic Press.

The four versions of the story are numbered for identification at the bottom of the second side. You will sort the data into the four cells for subsequent analyses: (1) victim attractive, defendant attractive; (2) victim unattractive, defendant attractive; (3) victim attractive, defendant unattractive; and (4) victim unattractive, defendant unattractive.

After you have collected the data, summarize your findings for all of the subjects interviewed by your classmates and yourself. Fill in the following table and turn it into your instructor.

Defendant (Sander)		Victim (Lowe)		
		Attractive	Unattractive	Total
Attractive	N =			
	M =			
	SD =			
Unattractive	N =			
	M =			
	SD =			
Total	N =			
	M =			
	SD =			

Your instructor will instruct you as to how to analyze the data. You can use a 2 (attractive vs. unattractive victim) x 2 (attractive vs. unattractive defendant) between-subjects analysis of variance. If the F-ratio is significant for the interaction, you can test specific preexperimental hypotheses by conducting specific comparisons between pairs of means. Separate ANOVAs or t-tests would be computed for each pair of means. As an additional analysis, you could determine the effect gender has upon the sentences. Interpret the results in light of the findings from Landy and Aronson (1969). Your instructor will direct you as to how to write up this study.

TOPIC 9: CONTROL

The goals of Topic 9 are to provide you with some experience in identifying variables that need to be controlled in experiments and conducting an experiment where practice effects need to be evaluated.

LABORATORY EXERCISE 1: IDENTIFYING UNCONTROLLED VARIABLES

The purpose of Laboratory Exercise I is to provide you with experience in identifying uncontrolled, or extraneous, variables.

Control is the most essential ingredient in experimentation. It is essential for the attainment of internal validity, the extent to which we can conclude that an independent variable produced an observed effect. In order to achieve control one attempts to eliminate any differential influence of extraneous variables. This means that a constant amount of the extraneous variable, assuming it cannot be eliminated, must exist at all levels of the independent variable for only then can one eliminate any differential influence. Therefore, the primary goal of control is to achieve constancy. However, before constancy can be achieved one must identify the relevant extraneous variables. This exercise gives you practice in this process. Before doing this, review the possible extraneous variables that may need to be controlled in a given experiment:

History. Any of the many events other than the independent variable that occur between a pre- and post-measurement of the dependent variable.

Maturation. Any of the many conditions internal to the individual that change as a function of the passage of time. These may involve biological and psychological processes.

Instrumentation. Any changes that occur as a function of measuring the dependent variable.

Statistical regression. Any change that can be attributed to the tendency of extremely high or low scores to regress toward the mean.

Selection. Any change due to the differential selection procedure used in placing subjects in various groups.

Mortality. Any change due to a differential subject loss from the various comparison groups.

Subject bias. Any change in performance that can be attributed to the subject's motives or attitudes.

Experimenter bias. Any change in the subject's performance that can be attributed to the experimenter.

Sequencing. Any change in the subject's performance that can be attributed to the fact that the subject participated in more than one treatment condition.

Interaction between selection and history, maturation or instrumentation. Any change in the subject's performance that can be attributed to one or more groups of subjects responding differentially to history, maturation, or instrumentation effects.

In this laboratory exercise you are presented with a series of examples (A-E) of experimental procedures. Your task is to identify the uncontrolled variable or variables in each research design. Indicate how you think these variables could have interacted with the independent variable(s) which were under investigation. Turn your answers into your instructor.

A. An investigator had hypothesized that one's store of general knowledge is related to our ability to acquire new information. The investigator randomly selected 15 subjects from the first, second, and third grades. He then had each group of subjects learn the same list of nonsense syllables by the method of serial anticipation and recorded the number of trials required to learn them to a criterion of one errorless trial. The results revealed that the first graders required an average of 25 trials to learn the list of nonsense syllables while the second graders required 20 trials and the third graders required 15 trials. The investigator concluded that the increase in general information provided by our educational system enhances a person's ability to acquire new information.

B. A pharmaceutical company developed a new drug which they thought would relieve depression and naturally was interested in supporting its investment. Consequently, the firm hired a team of researchers to investigate the potential effectiveness of the drug. These researchers identified a group of psychiatric patients who were experiencing chronic depression and randomly assigned half of the patients to the drug group and half of the patients to the placebo group. To avoid any possible confusion in administering the drug or placebo to the patients one psychiatric nurse was told that she was to administer the drug and another nurse was told that she was to administer the placebo. One month later the drug group had dramatically improved over the placebo group and the pharmaceutical company was elated and stated that it had developed a new effective anti-depressant drug.

C. A juvenile correctional institution had implemented a "buddy" system with its younger (6-12 year old) juveniles in an attempt to decrease the degree of deviancy exhibited by these children. Naturally, they wanted to determine if the "buddy" system was effective. To accomplish this they administered the "Behavioral Deviance" scale to their residents and selected, on the basis of this scale, the 15 percent of individuals who were the most extreme cases. These individuals were selected because if they could be helped by the "buddy" system then it should also help the less extreme cases. These extremely deviant individuals were then matched on the basis of age, sex, and race with a group of subjects who scored in the "normal" range on the "Behavioral Deviancy" scale. Both groups of subjects were exposed to the "buddy" system and it was hypothesized that the discrepancy between the scores of the two groups of subjects would decrease over time. After the "buddy" system had been implemented for two months the Behavioral Deviancy scale was again administered to the two groups of subjects. It was found that the average discrepancy between the scores of the two groups had decreased significantly. The administrators of the institution were elated and concluded that the "buddy" system was an effective means of reducing juvenile deviancy.

D. An investigator in a northeastern state attempted to isolate the effects of a statewide "crack-down" on speeding motorists. To accomplish this he recorded the number of traffic deaths and accidents between the months of November through April. To isolate the effects of the "crack-down" he had to make a comparison with an equivalent state that had not implemented such a "crack-down". Also, the comparison state had to be as equivalent as possible so he examined the records and identified a midwestern state who had a similar record of traffic accidents and deaths for the preceding two years. He then recorded the number of traffic deaths and accidents for both states for the six month period from November through April and found that there was no significant difference between the two states in number of deaths or accidents. He concluded that the crackdown was totally ineffective in having an influence on the two dependent variable measures he used.

E. An investigator hypothesized that subjects in a fear situation have a desire to affiliate with other individuals. To test his hypothesis the experimenter randomly assigned 50 subjects to either a high or low fear group. The low fear group was told that they would be shocked but that it would only be a small tingle which definitely would not hurt and some probably would not even feel it. The high fear group was told that they would be shocked and that the shock would be quite painful and may burn the skin but would not cause any permanent damage. After being told this, 10 subjects in the high fear group declined to continue to participate in the study. The experimenter released them as he was ethically bound to do and continued on with the experiment even though the number of subjects in the high fear condition was now reduced to 15. Each group of subjects was then told that they were to wait while the equipment was being prepared and that they could either wait in a room by themselves or with others. No difference was found in the extent to which the high and low fear subjects wanted to wait with others. The investigator concluded that fear was not related to affiliation.

LABORATORY EXERCISE 2: PRODUCTION OF VISUAL ILLUSION REVERSALS -- PRACTICE EFFECTS

The purpose of Laboratory Exercise 2 is to give you experience in evaluating the effect of practice upon the production of visual illusion reversals.

An illusion is a percept that is incorrect. The stimulus may be distorted in some way before it reaches our receptors, such as when you look at yourself in a distorting mirror at a fun house. Psychologists, for many years, have been greatly interested in perceptual illusions that arise in our own perceptual systems. They have asked such questions as how they occur, under what circumstances they occur, and why they occur.

Some of the best known visual illusions are the Muller-Lyer illusion and the Ponzo illusion, both of which you have probably seen in various psychology texts. Another set of illusions which have been of interest to psychologists involves reversible figures, such as the Necker Cube, which was devised in 1832 by the Swiss naturalist L. A. Necker, and the Schroeder Staircase. Look at the circle on the face of the cube below. As you concentrate upon the circle, you will probably notice that sometimes the circle appears as if it is on the front face and sometimes on the back face.

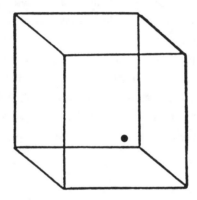

Because our perceptual system has no clues as to the correct pattern, it entertains both positions as possible and keeps going back and forth without settling on one solution. If more clues were given, then such a reversal would not occur (Gregory, 1977). Interestingly, there are individual differences in the number of reversals people report within a certain time period. It is suggested that the more one focuses upon the salient cues, the more likely an individual will report reversals (Power & Day, 1973).

This laboratory exercise will permit you to investigate (a) individual differences in the number of reversals seen in two reversible figures, and (b) the effect of practice upon the production of visual illusion reversals. The two figures, the Necker Cube and the Schroeder Staircase, are presented at the end of this exercise. Follow the instructions which are provided on the Response Sheet in this laboratory exercise.

After you have collected the data, compute the mean number of reversals you saw when looking at the Necker Cube and Schroeder Staircase and record it on the Response Sheet. On the following page you are provided a summary sheet for your class' scores. For each subject, write down the number of reversals reported for each of the 3 trials, as well as the mean number of reversals, for the Necker Cube and the Schroeder Staircase. Determine the mean number of reversals produced for each of the trials for all subjects. Plot a graph of the means across the three trials for the two reversal figures separately. Such a graph will indicate whether the means remain the same or vary. If they do vary, what is your hypothesis as to why they vary?

Next, compute a correlation (Pearson r or rho) between the mean number of reversals identified by the class members for the two reversal figures and determine whether it is significant. Evaluate whether practice had an effect upon the production of reversals by using either a repeated measures analysis of variance or a t-test for related (dependent) measures. Your instructor will direct you as to how to write up this study.

Why you think there are individual differences in the number of reversals reported for the two reversible figures? Research by Wallace and his students at Cleveland State University (Wallace, 1988; Wallace, Knight, & Garrett, 1976), as well as Crawford, Brown, and Moon (1993) found highly hypnotizable individuals reported more reversals than did low hypnotizable subjects. They hypothesized this is due to highly hypnotizable individuals' possessing better focused and sustained attention than lows. Coren and Porac (1987) have investigated the relationship between visuospatial abilities and susceptibility to certain visuo-geometric illusions. In a small sample of university students, Beer (1990) found no significant relationships between number of reversals for the Necker Cube or Schroeder Staircase and measures of visuospatial ability, curiosity, and intelligence. Knowledge or expectations may also influence the number of reversals reported. When uninformed about the reversibility of ambiguous figures, only about a third of subjects ever reversed; yet, when the same subjects were subsequently informed about the possibility of reversals, all reported reversals and did so frequently (Rock & Mitchener, 1992). These are interesting areas open to further research.

Visual Illusions References:

Beer, J. (1990). Correlations among ambiguous figures, curiosity, and spatial ability. *Perceptual and Motor Skills, 71*, 1188-1190.
Coren, S., & Porac, C. (1987). Individual differences in visual-geometric illusions: Predictions from measures of spatial cognitive abilities. *Perception and Psychophysics, 41*, 211-219.
Crawford, H. J., Brown, A. M., & Moon, C. E. (1993). Sustained attentional and disattentional abilities: Differences between low and highly hypnotizable individuals. *Journal of Abnormal Psychology, 102*, 534-543.
Gregory, R. L. (1977). *Eye and Brain: The psychology of seeing*. New York: McGraw-Hill.
Hoffman, D. D. (1983). The interpretation of visual illusions. *Scientific American, 249*, 154-162.
Power, R. P., & Day, R. H. (1973). Constancy and illusion of apparent direction of rotary motion in depth: Tests of a theory. *Perception and Psychophysics, 13*, 217-223.
Rock, I., & Mitchener, K. (1992). Further evidence of failure to reversal of ambiguous figures by uninformed subjects. *Perception, 21*, 39-45.
Shepard, R. N. (1990). *Mind sights: Original visual illusions, ambiguities, and other anomalies, with a commentary on the play of mind in perception and art*. New York: W. H. Freeman & Co.
Wallace, B. (1988). Hypnotic susceptibility, visual distraction, and reports of Necker cube apparent reversals. *Journal of General Psychology, 115*, 389-396.
Wallace, B., Knight, T. A., & Garrett, J. B. (1976). Hypnotic susceptibility and frequency reports to illusory stimuli. *Journal of Abnormal Psychology, 85*, 558-563.

RESPONSE SHEET: PRODUCTION OF VISUAL ILLUSION REVERSALS

Instructions:

Divide into pairs with your classmates. Each of you will serve, alternately, as subject or experimenter. Decide who will first be the subject.

The experimenter is to sit behind the subject out of his or her view. There should be no distractions within the subject's view. Place one of the two reversible figures in front of the subject, propped up so that it is comfortably placed 1 1/2 feet in front of the subject's face.

The experimenter will direct the subject as to when to start and stop each trial. There will be six trials, each lasting 60 seconds with 15 seconds between trials. During the 60 second trials the subject is to report "NOW" every time the figure reverses its figure-ground relationship. The experimenter is to record each reversal in the space below. The two figures, which are provided at the end of this laboratory exercise, are to be counterbalanced: one subject starts with the Necker Cube and the other starts with the Schroeder Staircase. Then the two figures are alternated with one another over the six trials so that there are three trials for each. Below the scoring area for each trial, the experimenter is to write down which figure is being presented on the line labeled "S".

To be consistent across subjects, the following procedure should be followed:

1. Subject sits 1 1/2 feet in front of figure and experimenter sits behind subject.

2. Experimenter asks subject "Are your ready?" and if he or she says yes, then the experiment should begin. There should be no extraneous talking!

3. The experimenter, using a stop watch, says "BEGIN" when the 60 second time period begins and "STOP" when it ends. The subject is to report, by saying "NOW", every time he or she sees a reversal in the figure-ground relationship during the 60 second time period.

4. After the trial is completed, the experimenter puts the other reversible figure in front of the subject and then starts over again after 15 seconds. This procedure continues for the six trials.

5. The experimenter then reverses roles with the subject.

TALLY HERE:

S: _____   S: _____   S: _____

Trial 1: _____   Trial 2: _____   Trial 3: _____

S: _____   S: _____   S: _____

Trial 4: _____   Trial 5. _____   Trial 6: _____

Mean of Necker Cube Reversals: _____   **Mean of Schroeder Staircase Reversals:** _____

SUMMARY SHEET: CLASS DATA

For each subject in your class, record the number of reversals reported during each trial, as well as the mean number of reversals, for the two reversible figures.

Subject #	Necker Cube				Schroeder Staircase			
	Trial 1	Trial 2	Trial 3	Mean	Trial 1	Trial 2	Trial 3	Mean
Mean =								
Standard Deviation =								

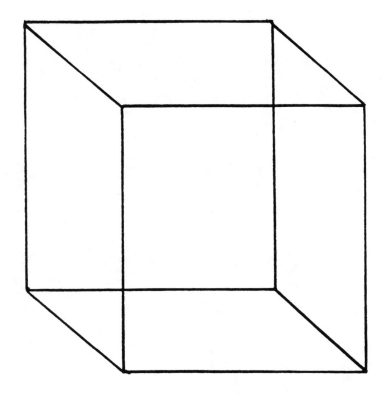

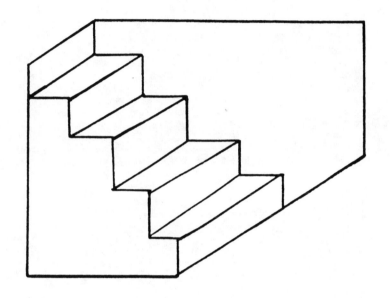

TOPIC 10: CONTROL TECHNIQUES

The goal of Topic 10 is to give you practice in identifying control techniques, in specifying the variables that are controlled by these techniques and in assigning subjects by various techniques. Before, beginning the exercises, it is important to review the control techniques that are commonly used.

Randomization. A control technique that provides each subject with an equal probability of being included in a given treatment group.

Matching. A variety of control techniques that accomplish the dual function of reducing error variance and at the same time controlling for the variable on which subjects are matched.

a. Matching by holding variables constant. Matching is achieved by having subjects in each experimental group have the same amount or type of the extraneous variable such as use of only males or only people with an IQ 120-125.

b. Matching by building the extraneous variable into the research design. In this case matching is achieved by having subjects with the same amount or type of an extraneous variable blocked or grouped together. The various blocks or groups would then form a series of levels of variation of that extraneous variable to be incorporated as an independent variable in the research design.

c. Yoked control. This is a control technique that eliminates the influence of the temporal relationship between an event and a response.

d. Equating subjects. Equating subjects means that the researcher identifies the variables on which he or she wants to match and then finds subjects that have the same amount or type of these variables.

Counterbalancing. A technique which controls for carryover and order sequencing effects.

Double blind placebo model. A control technique whereby neither the subject nor the experimenter is aware of which subjects took the treatment condition.

Deception. Involves providing the subject with a hypothesis unrelated to or orthogonal to the real hypothesis.

Disguised experiment. An experiment conducted in a context which does not reveal to the subjects that they are in an experiment.

Independent measurement of the dependent variable. The dependent variable is measured in a context totally removed from that of the independent variable manipulation.

The blind technique. The experimenter is unaware of subject's treatment condition.

Partial blind technique. The experimenter is unaware of the subject's treatment condition.

Automation. Removing the experimenter from the experimental context.

LABORATORY EXERCISE 1: IDENTIFYING CONTROL TECHNIQUES

The purpose of this laboratory exercise is to give you practice in identifying and evaluating the effectiveness of various control techniques in published studies.

(1) In your library, find a study in the experimental literature which used the double blind placebo model in a drug experiment with humans. If you are interested in a specific drug, you may look up its name in <u>Index Medicus</u> or <u>Psychological Abstracts</u>. On a separate sheet of paper, which you will hand into your instructor, answer the following.

   a. Give the complete reference.

   b. What is its purpose? What are the hypotheses?

   c. Describe the procedure of the experiment. What method was used to make the procedure double blind with a placebo?

   d. How effective was this control method? Were either the subjects or the experimenter aware of the treatment conditions? If so, how might this have affected the results?

   e. How would you improve on the control method used in this study?

(2) In your library, find a study in the experimental literature which used one of the following matching control techniques: a) matching by holding variables constant, (b) matching by building the extraneous variable into the research design, (c) yoked control, or (d) equating subjects on certain variables. On a separate sheet of paper, which you will hand into your instructor, answer the following:

   a. Give the complete reference.

   b. What is its purpose? What are the hypotheses?

   c. What method was used to match the subjects?

   d. How effective was this method? Was error variance reduced? Were the variables the experimenter thought he or she was controlling for actually controlled?

   e. What extraneous variables could still possibly confound the experiment?

   f. How would you improve on the control method used in this study?

LABORATORY EXERCISE 2: ASSIGNING SUBJECTS TO TREATMENT GROUPS

The purpose of this laboratory exercise is to give you practice in using various methods to assign subjects to treatment groups.

A graduate student in sports psychology at the University of Missouri proposed to answer this research question as part of her Master's thesis: What are the effects of viewing Olympic swimmers' training films on subsequent swimming skills?

In order to recruit subjects, she put a notice on the sign-up board asking for subjects who were nonswimmers (had never learned to swim) and interested in learning how to swim. Slots were open for a total of 40 subjects: 20 men and 20 women. Subjects were told to report at the same designated time to the women's gymnasium. At that time subjects were to be assigned to two "classes" of swimming training:

Class 1: Subjects would watch films of Olympic swimmers' training sessions before actually going into the pool for lessions.

Class 2: Subjects would go straight into the pool for lessons without seeing the films.

The dependent variable was "swimming ability". This was measured in two ways: (1) the obtained score on a short swimming skills knowledge test, and (2) the actual distance the subject could swim across the shallow end of the Olympic sized pool after a one hour training session.

The graduate student assigned subjects to the two classes on the basis of when they showed up for the session. The first 20 were assigned to Class #1 and the second 20 were assigned to Class #2. As can be seen in Table I, this selection procedure caused some potential problems: the subjects were not evenly matched. In Table I, the subjects are listed in the order in which they arrived. Their gender, weights, and IQs are also reported. As you can see, more women than men were the first to show up. In Class #I there were 6 men and 14 women, while in Class #2 there were 14 men and 6 women. Secondly, the mean weights of the two groups were significantly different: respectively, 135.2 and 149.95. Finally, the mean IQs were different: respectively, 111.2 and 103. Reflect for a moment as to why these differences might have occurred. These variables could possibly affect swimming ability. They are extraneous variables which need to be controlled.

Your task is to employ more appropriate "selection" procedures to form the two treatment groups ("classes") so that they are better equated on the variables under question. Specifically, you are to assign the subjects in Table 1 to the two groups, using each of the following three methods:

1. Random Assignment

2. Matching by precision control for (a) gender, (b) gender and weight, and (c) gender, weight, and IQ.

3. Matching building gender into the design.

Indepth information about these three methods are provided below.

METHOD 1: Straight Randomization

This technique is based on the premise that each subject has an equal chance of being selected for either experimental group (Class 1 or 2) in question. There are many ways this could be done. Here are two suggested ways.

A. Odd-Even Method
1. Go to a random numbers table, either in the Appendix of your experimental methods book or one provided by your instructor. Start anywhere you want in the table.
2. Reading one digit numbers in any direction, assign a random digit to each of the subjects in the pool (in this case, 1 - 40).
3. Make up a rule for randomization. In this case, subjects who have been assigned even numbers (0, 2, 4, 6, 8) will be place in treatment group 1, and those assigned odd numbers (1, 3, 5, 7, 9) will be placed in treatment group 2.
4. Once either group has been filled (20 subjects in this case) then whatever subjects are left over are assigned to the other group.

B. Direct Method
1. Go to a random numbers table. Start anywhere you want in the table.
2. Start running your finger in any direction looking for either a 1 (group 1) or a 2 (group 2). As soon as you come to a 1 or a 2, make the first subject go into that group.
3. Continuing from where you left off, again look for a 1 or a 2 for the second subject, and assign according to the number that comes up first (1 or 2).
4. Continue this way for all 40 subjects.
5. As soon as one group fills up, all other subjects must be placed in the unfilled group.

Whichever method is used, compare the characteristics of the randomly assigned groups to the groups selected by the graduate student. Answer the following:

1. How many males and females are in each of your groups?
2. What is the mean weight in each group?
3. What is the mean IQ in each group?

METHOD 2: Matching by Precision Control

If you wish to make absolutely sure that some of the traits are exactly equal in both groups, one approach you can take is matching by precision control.

A. Equating gender in each group
1. Go down the subject list. You find that the first subject is female.
2. Find another female to match the first female.
3. Assign one of these paired females to Group 1 and the other to Group 2, in a random fashion. Mark off each assigned female from your subject list.
4. Continue in this manner for the rest of the female subjects.
5. Now do the same thing for the male subjects, remembering to randomly assign the paired males to the two groups.

Now, check your groups. How many males and females are in each group? Have you used all 40 subjects?

B. Equating gender and weight in each group

1. Find the first subject and note this subject's "values" on the variables under consideration: gender and weight. As you will note, the first subject is female and weighs 107 pounds.
2. You are now to find another subject who matches this subject on gender and weight. If there is a perfect match, randomly assign these two subjects to the two groups. If there is no match, the first subject cannot be used and must be thrown out of the sample.
3. Continue in this manner for all 40 subjects. Remember to cross off subjects after they are assigned or thrown out.

Check your groups. How many subjects are in each group? How many males and females are in each group? What are their average weights? Did you use all subjects?

C. Equating gender, weight, and IQ in each group

Follow the same procedure as given above with all three variables. How many total subjects are used? What are the limitations of Methods B and C when you match by precision control?

METHOD 3: Matching by Building into Design

If you think a variable (such as gender, weight, and/or IQ) could have a significant effect on swimming ability, you can examine that effect along with the effect of the training film by matching and building these variables into your design. Matching by building a variable into the design will allow you to determine if that variable will have any effect.

If you wished to determine the effect of gender, as well as the film, on swimming ability, you would do the following:

1. Separate the males and females into two separate groups.
2. Randomly assign all males to groups 1 and 2.
3. Randomly assign all females to treatment groups 1 and 2.

Such an assignment would result in four treatment groups, which could be graphed in the following manner.

TREATMENT GROUPS

	Class #1	Class #2
Males	1	1
	.	.
	.	.
	n	n
GENDER		
Females	1	1
	.	.
	.	.
	n	n

TABLE 1

Subjects Listed in Chronological Order of Arrival

Name	Gender	Weight	IQ
First Twenty:			
1. Anna KannaFanna	F	109	110
2. Anita Newface	F	121	133
3. June Bugg	F	138	109
4. Wilbur Milburn	M	148	113
5. Illia Gitamate	F	136	101
6. Sam Dunkshot	M	195	148
7. Rick Shaw	M	185	130
8. Emma Knate	F	122	104
9. Ethyl Chloride	F	122	115
10. Mike Rofone	M	160	110
11. Fran Ennolly	F	113	122
12. Maye B. Knot	F	123	98
13. Jane Mundane	F	111	101
14. Polly Ester	F	134	120
15. I. C. Cream	M	167	94
16. Dee Leerious	F	147	96
17. Bob Frapples	M	152	112
18. Patti Kayck	F	105	100
19. Elenor Frozenfelt	F	120	103
20. Valerie Valerah	F	98	105
Second Twenty:			
21. Hal Itosis	M	163	82
22. Will Yadooit	M	180	110
23. Cora Lation	F	105	91
24. U. B. Trippin	M	167	106
25. Gail Warning	F	123	95
26. Barry A. Bone	M	210	117
27. Norm L. Curve	M	121	98
28. A. C. Deesey	M	135	102
29. Sharon Chairelyke	F	110	109
30. Oscar Meyer	M	139	131
31. Nick L. Beer	M	135	107
32. Buster Jaw	M	148	103
33. Jesse Minnitt	F	156	83
34. Cal Yafornia	M	178	112
35. Lena Onmie	F	119	85
36. Miles Standoffish	M	182	107
37. Elliot Mess	M	173	110
38. Rich O'Shay	M	152	105
39. Clara Nett	F	136	101
40. Johnny Stench	M	167	106

TOPIC 11: DESIGN

Topic 11 is concerned with the design of experiments. Several exercises will be provided to give you practice in actually designing experiments. Different research topics will be presented and your task will be to read the introductory material that describes the basic nature of the problem and then to design a study that investigates a component of this problem area. Since the introductory material will present only a brief introduction to the problem area, you would ordinarily not have sufficient knowledge to design an adequate experiment. To attempt to eliminate or at least minimize this deficit a variety of independent and dependent variables will be presented along with a typical procedure or procedures. Your task is to design a study using one or more of the independent and dependent variables specified. Your instructor will assign some or all of these exercises.

LABORATORY EXERCISE 1: DESIGNING AN OBESITY STUDY

During the past decade a great deal of scientific attention has been focused on the identification of factors relating to obesity. Within the general population, however, a dominant interest has naturally existed among those who are overweight. The overweight population seems to be continually seeking means for losing weight by means of one diet plan or another. Weight plans have proliferated, some of which have become faddish and captivated the interest of the overweight population. These weight reducing programs have been largely ineffective, judged by the fact that they do not enable the overweight person to reduce to and maintain a normal weight. A number of factors influence a person's food intake and weight gain. Research is directed toward the identification of these factors.

In the past, physiological psychologists have been interested in identifying the portion of the brain that controls eating behavior. Since such studies typically involve creating brain lesions, these studies are naturally not conducted on humans but on infrahumans such as rats, cats, and monkeys. These studies have found that bilateral lesions in the ventromedial nuclei of the hypothalamus will generally produce an animal that will eat prodigious amounts of food, producing a tremendous weight gain. The classic description of such an animal is that immediately following the operation, the animal staggers over to the food tray and begins voracious eating, which lasts for several weeks. Naturally there is an accompanying weight gain, which is labeled the "dynamic phase" of hyperphagia. Following this rapid weight gain a plateau is reached, the "static phase," at which point the animal's weight levels off and food intake drops to a level only slightly greater than that of the normal animal. During these two phases the lesioned animal remains inactive but emotional and irascible.

Such evidence would tend to suggest that a portion of the brain controls our eating behavior and maybe, in obese people, this control mechanism may have malfunctioned. However, other data suggest that there may be more involved. For example, if quinine is added to the food of a lesioned rat its food intake drops dramatically.

When turning to research on human subjects, we find additional suggestive data. In one of the earlier studies, Stunkard and Koch (1964) found that there was little relationship between gastric motility and self-report of hunger in obese persons whereas this relationship did exist in the normal subject. Such evidence indicates that the obese person does not label bodily states associated with food deprivation as hunger, whereas the normal person does. Such evidence suggests that there must be other factors that determine whether or not the obese person will ingest food. Your task is to design a study which investigates these other variables.

In designing this study assume the following setting:

All subjects are requested to report at the laboratory at 7:00 p.m. You had previously contacted them and asked them not to eat their evening meal prior to participating in the experiment. When the subject appears for the experiment you escorted him or her to a table and told the subject that you were conducting a study on taste.

Subjects are then presented with several kinds of crackers, almonds, or types of sandwiches and are asked to taste each kind and rate them on a rating scale. Subjects are also told to taste as many or as few crackers, almonds, or sandwiches of each type as they need in making their judgements. The important thing is that their rating be as accurate as possible.

Methodological Consideration

The primary consideration is the definition and measurement of obesity. As in the prior section dealing with independent variables, one must establish some criteria of what represents an obese person and what represents a normal person. This boils down to translating obesity into concrete operational terms. The most commonly used measures are as follows:

(1) Metropolitan Life Insurance tables of ideal or desired weight. Subjects are classified as obese if their weight deviates by 15% or more over what the tables labeled as the ideal or desired weights.

(2) Triceps skin fold thickness measurements. This is a measurement which determines the amount of subcutaneous fatty tissue in the body. This latter measure is perhaps better since it has been shown to be independent of the height or frame of the individual. A person's tricep skin-fold thickness is compared to a distribution of such scores (Seltzer and Mayer, 1965). In the past at least one study has classified obese people as individuals whose scores fell in the fourth quartile while subjects whose scores fell in the second quartile were classified as normal.

Possible Independent Variables

(1) Emotional Arousal - The psychosomatic hypothesis of obesity postulates that the obese eat in an attempt to cope with anxiety. In the laboratory this anxiety arousal could be created by threat of electrical shock.

(2) Food Cue Prominence - Schachter and his colleagues have emphasized the importance of seeing food. In other words, if food is highly visible it will operate as a cue to generate eating behavior. Therefore, to vary this variable one must conceive a way of varying the degree to which food cues are prominent as varying illumination or having the food wrapped.

(3) Taste - It has previously been found that rats with bilateral lesions in the ventromedial nuclei of the hypothalamus, after they have reached the static phase, will reduce their intake of food if it is laced with quinine. This suggests that taste may be a factor in the amount of food which is ingested by the obese. Therefore, to test the possibility of such a factor also affecting the behavior of humans one must vary the taste of the food given.

(4) Effort - It is possible that the obese, because they are heavier, may be less inclined to work for food than would their normal weight counterparts. To test this hypothesis it would be necessary to vary the difficulty in attaining the food.

(5) Gender - It is possible that male and female obese individuals respond differently in their eating patterns.

(6) Age - Obese children may respond differently than obese juveniles or obese adults.

Dependent Variable

In studies of obesity there is one primary dependent variable -- the amount eaten, whether it is number of crackers, nuts, or sandwiches, etc.

Laboratory Exercise 1: Design of the Research

Now that you have read the brief introductory material on obesity you are to conceptualize one or more research studies, to be specified by your instructor, that investigate one or more of the variables that have been specified. To accomplish this, you are to complete the following steps which correspond to steps that would actually be taken in the design of a study. Your instructor will go over these steps with you. Write it up on a separate sheet of paper and hand it into your instructor.

STEP 1. First Research Study: You are to first specify your research question and your scientific and null hypotheses.

Research question:

Scientific hypothesis:

Null hypothesis:

STEP 2. Design Specification: Now that you have specified your research question and your hypothesized outcome, you are to design an experiment that investigates this research using the APA Publication Manual guidelines.

Subjects:

Materials:

Procedure:

References on Obesity:

Abramson, E. E., & Wunderlich, R. A. (1972). Anxiety, fear and eating: A test of the psychosomatic concept of obesity. *Journal of Abnormal Psychology, 79*, 317-321.

Andrews, H. B., & Jones, S. (1990). Eating behaviour in obese women: A test of two hypotheses. *Australian Psychologist, 25*, 351-357.

Bowen, D. J., & Grunberg, N. E. (1987). Contributions of social psychology to the study of appetitive behaviors. *Journal of Applied Social Psychology, 17*, 622-640.

Ganley, R. M. (1989). Emotion and eating in obesity: A review of the literature. *International Journal of Eating Disorders, 8*, 343-361.

Leon, G. R., & Chamberlain, K. (1973). Comparison of daily eating habits and emotional state of overweight persons successful or unsuccessful in maintaining weight loss. *Journal of Consulting and Clinical Psychology, 41*, 108-115.

McKenna, R. J. (1972). Some effects of anxiety level and food cues on the eating behavior of obese and normal subjects: A comparison of the Schachterian and psychosomatic conceptions. *Journal of Personality and Social Psychology, 22*, 311-319.

Nisbett, R. E., & Gurwitz, S. B. (1970). Weight, sex, and the eating behavior of human newborns. *Journal of Comparative and Physiological Psychology, 73*, 245-253.

Reisenzein, R. (1983). The Schacter theory of emotion: Two decades later. *Psychological Bulletin, 94*, 239-264.

Rodin, J. (1981). Current status of the internal-external hypothesis for obesity. What went wrong? *American Psychologist, 36*, 361-372.

Rodin, J., Elias, M., Silverstein, L. R., & Wagner, A. (1988). Combined behavioral and pharmacologic treatment for obesity: Predictors of successful weight maintenance. *Journal of Consulting and Clinical Psychology, 56.* 399-404.

Rzewnicki, R., & Forgays, D. G. (1987). Recidivism and self-cure of smoking and obesity: An attempt to replicate. *American Psychologist, 42*, 97-100.

Schachter, S. (1971). Some extraordinary facts about obese humans and rats. *American Psychologist, 26*, 129-144.

Schachter, S., Goldman, R., & Gordon, A. (1968). Effects of fear, food deprivation, and obesity on eating. *Journal of Personality and Social Psychology, 10*, 91-97.

Singh, D. (1973). Role of response habits and cognitive factors in determination of behavior of obese humans. *Journal of Personality and Social Psychology, 27*, 220-238.

Spiegel, T. A., Shrager, E. E., & Stellar, E. (1989). Responses of lean and obese subjects to preloads, deprivation, and palatability. *Appetite, 13*, 45-69.

LABORATORY EXERCISE 2: DESIGNING AN INTERPERSONAL ATTRACTION STUDY

As we go through our everyday lives we encounter many people whom we like or to whom we are attracted. By the same token we encounter some people whom we dislike. Interpersonal attraction, therefore, refers to the evaluation of another person in a positive or negative way. Explanations for why we become attracted to certain individuals but not to others has been the focus of attention and discussion for centuries. The issue of attraction has been the focus of systematic research in psychology and sociology.

One may ask why it is important to study attraction? While many reasons can be given, such as its relationship to other social problems as marital failure, the primary objective of researchers investigating interpersonal attraction is much more basic. Interpersonal attraction is investigated in an attempt to identify the processes and empirical laws that operate to generate a positive or negative evaluation of others.

Interpersonal attraction was virtually an untouched research area until Moreno (1934) developed the sociometric technique. With his development of sociometry, serious attention was paid to the topic because now a method had been developed that would allow one to obtain some measure of attraction. Since Moreno's sociometric technique required each member of a group to specify with which other member or members he or she preferred to associate, investigators researched the variables that were related to one's popularity in groups. More recently other approaches have dominated the field. The development of these additional approaches has been necessary to provide further impetus to the exploration of attraction. One of the paradigms that has been developed is Byrne's (1971) attraction paradigm. Attention will be focused on this approach since it has been extremely successful in enabling investigators to research a variety of antecedent issues which have increased our knowledge of the conditions surrounding attraction.

The attraction paradigm, which is advocated by Byrne, is a reinforcement paradigm or a conditioning model. Byrne proposed that any stimulus with reinforcement properties could function as an unconditioned stimulus in producing an affective response or a response following along the pleasant-unpleasant continuum. Any discriminable stimulus, such as a person, that is temporally associated with this unconditioned stimulus would then be conditioned to evoke the same implicit affective response as did the unconditioned stimulus. Since Byrne and his colleagues demonstrated that attitude statements can operate as unconditioned stimuli eliciting either positive or negative affective responses, a person (CS) who is associated with positive or negative attitudes (UCS) would through conditioning come to also elicit these positive or negative affective responses. Since these affective responses mediate evaluative responses such as like, dislike, and hate, a person would, according to Byrne and his colleagues, come to like or dislike another based on the attitudes he or she associated with that person.

To investigate the validity of this paradigm, Byrne and his colleagues devised the following procedures: During the first part of the semester subjects are given the Survey of Attitudes (a 56-item attitude scale, developed in the 1960s, located at the end of this exercise). These subjects are then randomly assigned to the designated experimental groups. A second bogus stranger then supposedly has taken the same attitude scale. This bogus stranger's responses are patterned by the experimenter so they will be similar or dissimilar to those expressed by the subject. There are several different ways in which these fake patterns of responses can be generated. Table 1 illustrates the possible method that can be used.

Table 1: Different Faking Patterns for Similar and Dissimilar Item Responses by Bogus Strangers[1]

Subjects' Responses	Identity -	Mirror	Moderate Discrepancy		Constant Discrepancy	
	Similar	Dissimilar	Similar	Dissimilar	Similar	Dissimilar
1	1	6	2	5	2	4
2	2	5	3	4	1 or 3	5
3	3	4	2	5	2	6
4	4	3	5	2	5	1
5	5	2	4	3	4 or 6	2
6	6	1	5	2	5	3

[1]From Bryne, D. (1969). Attitudes and attraction. In L. Berkowitz (ed.), *Advances in Experimental Social Psychology*. New York: Academic Press.

Consequently, if you wanted to use the moderate discrepancy pattern in creating a similar bogus subject you would identify the subject's response (e.g., 4) and locate it in the first column. Then move across this row until you reach the moderate discrepancy/similar column. The number in that column (5) would represent the bogus subject's response. This procedure would be followed until you had provided a response for all items in the bogus subject's Survey of Attitudes scale. The actual subject's Survey of Attitudes scale along with his or her paired bogus subject's scale would then be returned to the subject with the following instructions. (The background information section is cut off with a scissors. Also subjects are to complete the Interpersonal Judgement Scale after forming an opinion of the bogus person since this represents the dependent variable measure).

Earlier this semester, you filled out an attitude questionnaire called the Survey of Attitudes which dealt with a series of issues. One purpose was to learn something about student attitudes, but a second purpose was to determine the extent to which one person can form valid judgements about another person just by knowing a few of his attitudes. Last semester we carried out other studies of this sort. Students wrote down several sorts of information about themselves, their names were removed, and this information was given to other students. The task was to form an opinion about the stranger's intelligence, knowledge of current events, morality, and adjustment just on the basis of knowing a few bits of information about the person's past and present life. We found that students could guess these things with better than chance accuracy. So, this study is an extension of the previous one, and a major change has been introduced. Instead of information about the other person's life, you will be shown his or her attitudes on 26 specific issues. The background information was removed from each of these scales. Each of you will receive the attitude scale of another student. All I can guarantee is that this person is the same sex as yourself, to the best of our knowledge you do not know the person whose attitude scale you will receive, and it is not someone in the same psychology class as yourself. Please read the person's answers carefully and try to form an opinion about him or her. As soon as you have studied each of the attitudes, fill out the Interpersonal Judgement Scale and indicate your best guess as to this person's intelligence, knowledge of current events, morality, and adjustment. Also, indicate how much you think you would like to work with this person as partners in an experiment (Bryne, 1971, p. 51).

Possible Independent Variables

A. Proportion of Similar Attitudes. Since Byrne and his colleagues have advocated that positive attitudes elicit positive affective responses such as liking, a logical extension of this is that the greater the proportion of similar attitudes expressed by the bogus stranger, the greater the expressed liking for that stranger. To test the possibility one would have to vary the proportion of similar attitudes expressed by the bogus stranger to various groups of subjects.

B. Topic Importance. One of the issues which can be raised when one discusses the attitude attraction field is the issue of the importance of the topic on which subjects agree or disagree. It could be hypothesized that agreement or disagreement on important topics leads to greater liking or disliking than does agreement or disagreement on unimportant items. To investigate this possibility the items in the 56 item attitude scale must first be scaled for importance. Fortunately this has already been accomplished, allowing this hypothesis to be tested. The 14 most important items are items 5, 6, 10,

14, 17, 19, 25, 27, 31, 33, 34, 43, 46, 48. The next most important items are 12, 21, 45, 52, 55, 28, 22, 39, 51, 40, 3, 35, 56, 24; the 14 next-to-least important items are 53, 47, 8, 42, 13, 23, 16, 30, 44, 7, 29, 18, 1, 41; and the 14 least important items are 9, 50, 32, 15, 54, 37, 26, 36, 4, 38, 2, 11, 49, 20.

C. Sequential Presentation of Attitudinal Stimuli. In the field of impression formation one area of active research is on the impact of first impressions. This research has revealed that first impressions are often inaccurate and that we revise them as we continue interaction. A similar hypothesis could be stated about the relationship between attitudes and attraction. In most interpersonal encounters one hears attitudes and opinions that are both similar and dissimilar to one's own. Does the order in which these similar and dissimilar attitudes appear influence one's attraction?

D. Prestige or Status of the Bogus Stranger. In our daily encounter with others we come across others who exhibit greater or lesser prestige in our eyes. Prior sociometric studies have reported a positive relationship between prestige and friendship choices. One could then ask if this prestige factor operated over and above attitude similarity in producing attraction. If it did, then attraction from a high prestige person should be greater than that expected from a member of a peer group. In this context one might also inquire as to the effects generated by a prestigious person exhibiting attitude dissimilarity. It is very possible that such a person could elicit the greatest negative impact. Prestige could be altered by attributing to the bogus stranger a given occupational status, military rank, etc.

E. Gender. This is an obvious variable; however, it is one that may produce an interaction. For example, it is possible that similarity in attitude is more important in forming friendships with individuals of the same gender. Opposite-gendered friendships may be less dependent upon attitude similarity but more dependent upon factors such as physical attractiveness.

F. Emotional Disturbance. Byrne's reinforcement paradigm reveals hypotheses that attitude similarity should lead to attraction. However, what would be the relationship if the other or bogus stranger possessed attributes that were undesirable to us, such as emotional disturbance? Would such an undesirable attribute operate to reduce the effectiveness of attitude similarity? It is possible that it might since an attribute would make similarity unpleasant or even threatening.

G. Race. It has been suggested in the literature that racial prejudice is actually reducible to an assumption that individuals of a different race possess different beliefs. If this is so, then knowledge of belief similarity should override any racial prejudice that would exist. Within the Byrne paradigm it should be possible to determine if racial prejudice is a function of race or belief similarity.

Dependent Variable

Although there are a variety of dependent variables that have been identified for use in attraction research, the one that is consistently identified and used with the Byrne Attraction Paradigm is contained within the Interpersonal Judgement Scale (located at the end of this exercise). Although this scale requires each person to rate the bogus stranger on 7 items, only two of these items are used for assessment of attraction. These two items, "Would you like or dislike this person?" and, "Would you like or dislike working with this person?" are summed to provide a single measure of attraction ranging from a score of 2 (most negative) to a score of 14 (most positive). These two items are embedded in the context of the other 5 items to give the instructions concerning interpersonal judgement credibility.

Laboratory Exercise 2: Design of the Experiment

Now that you have read the introductory material concerning attraction, as well as the potential independent variables that could be used, you are to construct one or more research studies (to be specified by your laboratory instructor) that would investigate one or more of the specified independent variables. To accomplish this you are to complete the following steps.

STEP 1. First Research Study: You are to first specify your research question and your scientific or null hypothesis.

Research question:

Scientific hypothesis:

Null hypothesis:

STEP 2. Design Specification: Now that you have specified your research question and your hypothesized outcome, you are to design an experiment that investigates this research question using the APA Publication Manual guidelines.

Subjects:

Materials:

Procedure:

References on Attraction:

Byrne, D. (1969). Attitudes and attraction. In L. Berkowitz (Ed.), *Advances in Experimental Social Psychology*. New York: Academic Press, Vol. 5.

Byrne, D. (1971). *The Attraction Paradigm*. New York: Academic Press.

Condon, J. W., & Crano, W. D. (1988). Inferred evaluation and the relation between attitude similarity and interpersonal attraction. *Journal of Personality and Social Psychology, 54*, 789-797.

Duck, S. (Ed.) (1977). *Theory and Practice in Interpersonal Attraction*. New York: Academic Press.

Grover, S. L., & Brockner, J. (1989). Empathy and the relationship between attitudinal similarity and attraction. *Journal of Research in Personality, 23*, 469-479.

Jamieson, D. W., Lyndon, J. E., & Zanna, M. P. (1987). Attitude and activity preference similarity: Differential bases of interpersonal attraction for low and high self-monitors. Special Issue: Integrating personality and social psychology. *Journal of Personality and Social Psychology, 53*, 1052-1060.

Smeaton, G., Byrne, D., & Murnen, S. K. (1989). The repulsion hypothesis revisited: Similarity irrelevance or dissimilarity bias? *Journal of Personality and Social Psychology, 56*, 54-59.

LABORATORY EXERCISE 3: CARRYING OUT AN INTERPERSONAL ATTRACTION EXPERIMENT

Now that you have designed an experiment in Laboratory Exercise 2, your instructor will select one of the proposed experiments to be conduced using the class as subjects. The experimenter will be the student whose design has been selected. The primary criteria for selection will be adequacy of the design and its ability to be used within the context of the class. In carrying out the study, most of you will be subjects. However, you are not naive in the same way as most subjects are since you have read a little about interpersonal attraction and you designed a study related to it. Therefore, when participating as a subject you are going to have to role play and act as much like a naive subject as you can. This is extremely necessary for the data to be as uncontaminated as possible.

Experiment Proper. Your student experimenter and instructor will instruct you in the tasks you are to complete in your capacity as a subject. The Attitude Scale and Interpersonal Judgment Scale are provided on subsequent pages.[2]

Data Analysis. Now that you have served as a subject you are to obtain the data and analyze it as though you were the experimenter. Your instructor will give you the data in addition to explaining the design of the experiment. At this point you must ask any questions you think are necessary to clarify the experiment and to insure that you know what the experiment was attempting to accomplish. To assist in this regard you should record the research question null or scientific hypothesis. Once you have laid out your design you can record the raw data in the appropriate categories.

Research question:

Null hypothesis:

Scientific hypothesis:

Now that you have the raw data you are to analyze the data with the help of your laboratory instructor. When you complete the statistical analysis you are to write up the results as close as possible to APA format. Additional information on the APA format can be obtained from your experimental psychology textbook or from the APA *Publication Manual*.

---

[2]D. Byrne, *Attraction Paradigm*, pp. 416-427. New York: Academic Press, 1971, with permission.

Fifty-Six Item Attitude Scale

Survey of Attitudes

Name: _____ Psy.: _____ Sec.: _____ Date: _____

Age: _____ Sex: _____ Class: Fr. _____ Soph. _____ Jr. _____ Sr. _____

Hometown: _____

1. Fraternities and Sororities (check one)

_____ I am very much against fraternities and sororities as they usually function.
_____ I am against fraternities and sororities as they usually function.
_____ To a slight degree, I am against fraternities and sororities as they usually function.
_____ To a slight degree, I am in favor of fraternities and sororities as they usually function.
_____ I am in favor of fraternities and sororities as they usually function.
_____ I am very much in favor of fraternities and sororities as they usually function.

2. Western Movies and Television (check one)

_____ I enjoy western movies and television programs very much.
_____ I enjoy western movies and television programs.
_____ I enjoy western movies and television programs to a slight degree.
_____ I dislike western movies and television programs to a slight degree.
_____ I dislike western movies and television programs.
_____ I dislike western movies and television programs very much.

3. Undergraduates Getting Married (check one)

_____ In general, I am very much in favor of undergraduates getting married.
_____ In general, I am in favor of undergraduates getting married.
_____ In general, I am mildly in favor of undergraduates getting married.
_____ In general, I am mildly against undergraduates getting married.
_____ In general, I am against undergraduates getting married.
_____ In general, I am very much against undergraduates getting married.

4. Situation Comedies (check one)

_____ I dislike situation comedies very much.
_____ I dislike situation comedies.
_____ I dislike situation comedies to a slight degree.
_____ I enjoy situation comedies to a slight degree.
_____ I enjoy situation comedies.
_____ I enjoy situation comedies very much.

5. Belief in God (check one)

_____ I strongly believe that there is a God.
_____ I believe that there is a God.
_____ I feel that perhaps there is a God.
_____ I feel that perhaps there is no God.
_____ I believe that there is no God.
_____ I strongly believe that there is no God.

6. Professors and Student Needs (check one)

_____ I feel that university professors are completely indifferent to student needs.
_____ I feel that university professors are indifferent to student needs.
_____ I feel that university professors are slightly indifferent to student needs.
_____ I feel that university professors are slightly concerned about student needs.
_____ I feel that university professors are concerned about student needs.
_____ I feel that university professors are very much concerned about student needs.

7. Draft (check one)

_____ I am very much in favor of the draft.
_____ I am in favor of the draft.
_____ I am mildly in favor of the draft.
_____ I am mildly opposed to the draft.
_____ I am opposed to the draft.
_____ I am very much opposed to the draft.

8. Necking and Petting (check one)

_____ In general, I am very much against necking and petting among couples in college.
_____ In general, I am against necking and petting among couples in college.
_____ In general, I am mildly against necking and petting among couples in college.
_____ In general, I am mildly in favor of necking and petting among couples in college.
_____ In general, I am in favor of necking and petting among couples in college.
_____ In general, I am very much in favor of necking and petting among couples in college.

9. Smoking (check one)

_____ In general, I am very much in favor of smoking.
_____ In general, I am in favor of smoking.
_____ In general, I am mildly in favor of smoking.
_____ In general, I am mildly against smoking.
_____ In general, I am against smoking.
_____ In general, I am very much against smoking.

10. Integration in Public Schools (check one)

_____ Racial integration in public schools is a mistake, and I am very much against it.
_____ Racial integration in public schools is a mistake, and I am against it.
_____ Racial integration in public schools is a good plan, and I am mildly against it.
_____ Racial integration in public schools is a good plan, and I am mildly in favor of it.
_____ Racial integration in public schools is a good plan, and I am in favor of it.
_____ Racial integration in public schools is a good plan, and I am very much in favor of it.

11. **Comedians Who Use Satire** (check one)

_____ I very much enjoy comedians who use satire.
_____ I enjoy comedians who use satire.
_____ I mildly enjoy comedians who use satire.
_____ I mildly dislike comedians who use satire.
_____ I dislike comedians who use satire.
_____ I very much dislike comedians who use satire.

12. **Acting on Impulse vs. Careful Consideration of Alternatives** (check one)

_____ I feel that it is better if people always act on impulse.
_____ I feel that it is better if people usually act on impulse.
_____ I feel that it is better if people often act on impulse.
_____ I feel that it is better if people often engage in a careful consideration of alternatives.
_____ I feel that it is better if people usually engage in a careful consideration of alternatives.
_____ I feel that it is better if people always engage in a careful consideration of alternatives.

13. **Social Aspects of College Life** (check one)

_____ In general, I am very much against an emphasis on the social aspects of college life.
_____ In general, I am against an emphasis on the social aspects of college life.
_____ In general, I am mildly against an emphasis on the social aspects of college life.
_____ In general, I am mildly in favor of an emphasis on the social aspects of college life.
_____ In general, I am in favor of an emphasis on the social aspects of college life.
_____ In general, I am very much in favor of an emphasis on the social aspects of college life.

14. **Birth Control** (check one)

_____ I am very much in favor of most birth control techniques.
_____ I am in favor of most birth control techniques.
_____ I am mildly in favor of most birth control techniques.
_____ I am mildly opposed to most birth control techniques.
_____ I am opposed to most birth control techniques.
_____ I am very much opposed to most birth control techniques.

15. **Classical Music** (check one)

_____ I dislike classical music very much.
_____ I dislike classical music.
_____ I dislike classical music to a slight degree.
_____ I enjoy classical music to a slight degree.
_____ I enjoy classical music.
_____ I enjoy classical music very much.

16. **Drinking** (check one)

_____ In general, I am very much in favor of college students drinking alcoholic beverages.
_____ In general, I am in favor of college students drinking alcoholic beverages.
_____ In general, I am mildly in favor of college students drinking alcoholic beverages.
_____ In general, I am mildly opposed to college students drinking alcoholic beverages.
_____ In general, I am opposed to college students drinking alcoholic beverages.
_____ In general, I am very much opposed to college students drinking alcoholic beverages.

17. **American Way of Life** (check one)

_____ I strongly believe that the American way of life is not the best.
_____ I believe that the American way of life is not the best.
_____ I feel that perhaps the American way of life is not the best.
_____ I feel that perhaps the American way of life is the best.
_____ I believe that the American way of life is the best.
_____ I strongly believe that the American way of life is the best.

18. **Sports** (check one)

_____ I enjoy sports very much.
_____ I enjoy sports.
_____ I enjoy sports to a slight degree.
_____ I dislike sports to a slight degree.
_____ I dislike sports.
_____ I dislike sports very much.

19. **Premarital Sex Relations** (check one)

_____ In general, I am very much opposed to premarital sex relations.
_____ In general, I am opposed to premarital sex relations.
_____ In general, I am mildly opposed to premarital sex relations.
_____ In general, I am mildly in favor of premarital sex relations.
_____ In general, I am in favor of premarital sex relations.
_____ In general, I am very much in favor of premarital sex relations.

20. **Science Fiction** (check one)

_____ I enjoy science fiction very much.
_____ I enjoy science fiction.
_____ I enjoy science fiction to a slight degree.
_____ I dislike science fiction to a slight degree.
_____ I dislike science fiction.
_____ I dislike science fiction very much.

21. **Money** (check one)

_____ I strongly believe that money is not one of the most important goals in life.
_____ I believe that money is not one of the most important goals in life.
_____ I feel that perhaps money is not one of the most important goals in life.
_____ I feel that perhaps moneys is one of the most important goals in life.
_____ I believe that money is one of the most important goals in life.
_____ I strongly believe that money is one of the most important goals in life.

22. **Grades** (check one)

_____ I am very much in favor of the university grading system as it now exists.
_____ I am in favor of the university grading system as it now exists.
_____ I am slightly in favor of the university grading system as it now exists.
_____ I am slightly opposed to the university grading system as it now exists.
_____ I am opposed to the university grading system as it now exists.
_____ I am very much opposed to the university grading system as it now exists.

23. **Political Parties (check one)**

_____ I am a strong supporter of the Democratic party.
_____ I prefer the Democratic party.
_____ I have a slight preference to the Democratic party.
_____ I have a slight preference to the Republican party.
_____ I prefer the Republican party.
_____ I am a strong supporter of the Republican party.

24. **Group Opinion (check one)**

_____ I feel that people should always ignore group opinion if they disagree with it.
_____ I feel that people should usually ignore group opinion if they disagree with it.
_____ I feel that people should often ignore group opinion if they disagree with it.
_____ I feel that people should often go along with group opinion even if they disagree with it.
_____ I feel that people should usually go along with group opinion even if they disagree with it.
_____ I feel that people should always go along with group opinion even if they disagree with it.

25. **One True Religion (check one)**

_____ I strongly believe that my church represents the one true religion.
_____ I believe that my church represents the one true religion.
_____ I feel that probably my church represents the one true religion.
_____ I feel that probably no church represents the one true religion.
_____ I believe that no church represents the one true religion.
_____ I strongly believe that no church represents the one true religion.

26. **Musical Comedies (check one)**

_____ I dislike musical comedies very much.
_____ I dislike musical comedies.
_____ I dislike musical comedies to a slight degree.
_____ I enjoy musical comedies to a slight degree.
_____ I enjoy musical comedies.
_____ I enjoy musical comedies very much.

27. **Preparedness for War (check one)**

_____ I strongly believe that preparedness for war will not tend to precipitate war.
_____ I believe that preparedness for war will not tend to precipitate war.
_____ I feel that perhaps preparedness for war will not tend to precipitate war.
_____ I feel that perhaps preparedness for war will tend to precipitate war.
_____ I believe that preparedness for war will tend to precipitate war.
_____ I strongly believe that preparedness for war will tend to precipitate war.

28. **Welfare Legislation (check one)**

_____ I am very much opposed to increased welfare legislation.
_____ I am opposed to increased welfare legislation.
_____ I am mildly opposed to increased welfare legislation.
_____ I am mildly in favor of increased welfare legislation.
_____ I am in favor of increased welfare legislation.
_____ I am very much in favor of increased welfare legislation.

29. **Creative Work** (check one)

_____ I enjoy doing creative work very much.

_____ I enjoy doing creative work.

_____ I enjoy doing creative work to a slight degree.

_____ I dislike doing creative work to a slight degree.

_____ I dislike doing creative work.

_____ I dislike doing creative work very much.

30. **Dating** (check one)

_____ I strongly believe that girls should be allowed to date before they are in high school.

_____ I believe that girls should be allowed to date before they are in high school.

_____ I feel that perhaps girls should be allowed to date before they are in high school.

_____ I feel that perhaps girls should not be allowed to date before they are in high school.

_____ I believe that girls should not be allowed to date before they are in high school.

_____ I strongly believe that girls should not be allowed to date before they are in high school.

31. **Red China and the U.N.**[3] (check one)

_____ I strongly believe that Red China should not be admitted to the U.N.

_____ I believe that Red China should not be admitted to the U.N.

_____ I feel that perhaps Red China should not be admitted to the U.N.

_____ I feel that perhaps Red China should be admitted to the U.N.

_____ I believe that Red China should be admitted to the U.N.

_____ I strongly believe that Red China should be admitted to the U.N.

32. **Novels** (check one)

_____ I dislike reading novels very much.

_____ I dislike reading novels.

_____ I dislike reading novels to a slight degree.

_____ I enjoy reading novels to a slight degree.

_____ I enjoy reading novels.

_____ I enjoy reading novels very much.

33. **Socialized Medicine** (check one)

_____ I am very much opposed to socialized medicine as it operates in Great Britain.

_____ I am opposed to socialized medicine as it operates in Great Britain.

_____ I am mildly opposed to socialized medicine as it operates in Great Britain.

_____ I am mildly in favor of socialized medicine as it operates in Great Britain.

_____ I am in favor of socialized medicine as it operates in Great Britain.

_____ I am very much in favor of socialized medicine as it operates in Great Britain.

34. **War** (check one)

_____ I strongly feel that war is sometimes necessary to solve world problems.

_____ I feel that war is sometimes necessary to solve world problems.

_____ I feel that perhaps war is sometimes necessary to solve world problems.

_____ I feel that perhaps war is never necessary to solve world problems.

_____ I feel that war is never necessary to solve world problems.

_____ I strongly feel that war is never necessary to solve world problems.

Authors note: To update this question, you may wish to change "Red China" to "People's Republic of China" and use a more current issue, such as that of the United States' diplomatic relations with China.

35. State Income Tax (check one)

_____ I am very much opposed to a state income tax.
_____ I am opposed to a state income tax.
_____ I am mildly opposed to a state income tax.
_____ I am mildly in favor of a state income tax.
_____ I am in favor of a state income tax.
_____ I am very much in favor to a state income tax.

36. Tipping (check one)

_____ I am very much opposed to the custom of tipping.
_____ I am opposed to the custom of tipping.
_____ I am mildly opposed to the custom of tipping.
_____ I am mildly in favor of the custom of tipping.
_____ I am in favor of the custom of tipping.
_____ I am very much in favor of the custom of tipping.

37. Pets (check one)

_____ I enjoy keeping pets very much.
_____ I enjoy keeping pets.
_____ I enjoy keeping pets to a slight degree.
_____ I dislike keeping pets to a slight degree.
_____ I dislike keeping pets.
_____ I dislike keeping pets very much.

38. Foreign Movies (check one)

_____ I enjoy foreign movies very much.
_____ I enjoy foreign movies.
_____ I enjoy foreign movies to a slight degree.
_____ I dislike foreign movies to a slight degree.
_____ I dislike foreign movies.
_____ I dislike foreign movies very much.

39. Strict Discipline (check one)

_____ I am very much against strict disciplining of children.
_____ I am against strict disciplining of children.
_____ I am mildly against strict disciplining of children.
_____ I am mildly in favor of strict disciplining of children.
_____ I am in favor of strict disciplining of children.
_____ I am very much in favor of strict disciplining of children.

40. Financial Help From Parents (check one)

_____ I strongly believe that parents should provide financial help to young married couples.
_____ I believe that parents should provide financial help to young married couples.
_____ I feel that perhaps parents should provide financial help to young married couples.
_____ I feel that perhaps parents should not provide financial help to young married couples.
_____ I believe that parents should not provide financial help to young married couples.
_____ I strongly believe that parents should not provide financial help to young married couples.

41. **Freshman Having Cars on Campus** (check one)

_____ I am very much in favor of freshmen being allowed to have cars on campus.

_____ I am in favor of freshmen being allowed to have cars on campus.

_____ I am in favor of freshmen being allowed to have cars on campus to a slight degree.

_____ I am against freshmen being allowed to have cars on campus to a slight degree.

_____ I am against freshmen being allowed to have cars on campus.

_____ I am very much against freshmen being allowed to have cars on campus.

42. **Foreign Language** (check one)

_____ I am very much in favor of requiring students to learn a foreign language.

_____ I am in favor of requiring students to learn a foreign language.

_____ I am mildly in favor of requiring students to learn a foreign language.

_____ I am mildly opposed to requiring students to learn a foreign language.

_____ I am opposed to requiring students to learn a foreign language.

_____ I am very much opposed to requiring students to learn a foreign language.

43. **College Education** (check one)

_____ I strongly believe it is very important for a person to have a college education in order to be successful.

_____ I believe it is very important for a person to have a college education in order to be successful.

_____ I believe that perhaps it is very important for a person to have a college education in order to be successful.

_____ I believe that perhaps it is not very important for a person to have a college education in order to be successful.

_____ I believe that it is not very important for a person to have a college education in order to be successful.

_____ I strongly believe that it is not very important for a person to have a college education in order to be successful.

44. **Fresh Air and Exercise** (check one)

_____ I strongly believe that fresh air and daily exercise are not important.

_____ I believe that fresh air and daily exercise are not important.

_____ I feel that probably fresh air and daily exercise are not important.

_____ I feel that probably fresh air and daily exercise are important.

_____ I believe fresh air and daily exercise are important.

_____ I strongly believe that fresh air and daily exercise are important.

45. **Discipline of Children** (check one)

_____ I strongly believe that the father should discipline the children in the family.

_____ I believe that the father should discipline the children in the family.

_____ I feel that perhaps the father should discipline the children in the family.

_____ I feel that perhaps the mother should discipline the children in the family.

_____ I believe that the mother should discipline the children in the family.

_____ I strongly believe that the mother should discipline the children in the family.

46. **Nuclear Arms Race** (check one)

_____ I am very much opposed to the federal government's buildup of nuclear arms.

_____ I am opposed to the federal government's buildup of nuclear arms.

_____ I am mildly opposed to the federal government's buildup of nuclear arms.

_____ I am mildly in favor of the federal government's buildup of nuclear arms.

_____ I am in favor of the federal government's buildup of nuclear arms.

_____ I am very much in favor of the federal government's buildup of nuclear arms.

47. Community Bomb Shelters (check one)

_____ I strongly believe that the federal government should provide community bomb shelters.
_____ I believe that the federal government should provide community bomb shelters.
_____ I feel that perhaps the federal government should provide community bomb shelters.
_____ I feel that perhaps individuals should provide their own bomb shelters.
_____ I believe that individuals should provide their own bomb shelters.
_____ I strongly believe that individuals should provide their own bomb shelters.

48. Divorce (check one)

_____ I am very much opposed to divorce.
_____ I am opposed to divorce.
_____ I am mildly opposed to divorce.
_____ I am mildly in favor of divorce.
_____ I am in favor of divorce.
_____ I am very much in favor of divorce.

49. Gardening (check one)

_____ I enjoy gardening very much.
_____ I enjoy gardening.
_____ I enjoy gardening to a slight degree.
_____ I dislike gardening to a slight degree.
_____ I dislike gardening.
_____ I dislike gardening very much.

50. Dancing (check one)

_____ I enjoy dancing very much.
_____ I enjoy dancing.
_____ I enjoy dancing to a slight degree.
_____ I dislike dancing to a slight degree.
_____ I dislike dancing.
_____ I dislike dancing very much.

51. A Catholic President (check one)

_____ I am very much in favor of a Catholic being elected president.
_____ I am in favor of a Catholic being elected president.
_____ I am mildly in favor of a Catholic being elected president.
_____ I am mildly against a Catholic being elected president.
_____ I am against a Catholic being elected president.
_____ I am very much against a Catholic being elected president.

52. Women in Today's Society (check one)

_____ I strongly believe that women are not taking too aggressive a role in society today.
_____ I believe that women are not taking too aggressive a role in society today.
_____ I feel that perhaps women are not taking too aggressive a role in society today.
_____ I feel that perhaps women are taking too aggressive a role in society today.
_____ I believe that women are taking too aggressive a role in society today.
_____ I strongly believe that women are taking too aggressive a role in society today.

53. **Family Finances** (check one)

_____ I strongly believe that the man in the family should handle the finances.
_____ I believe that the man in the family should handle the finances.
_____ I feel that perhaps the man in the family should handle the finances.
_____ I feel that perhaps the woman in the family should handle the finances.
_____ I believe that the woman in the family should handle the finances.
_____ I strongly believe that the woman in the family should handle the finances.

54. **Exhibitions of Modern Art** (check one)

_____ I dislike looking at exhibitions of modern art very much.
_____ I dislike looking at exhibitions of modern art.
_____ I dislike looking at exhibitions of modern art to a slight degree.
_____ I enjoy looking at exhibitions of modern art to a slight degree.
_____ I enjoy looking at exhibitions of modern art.
_____ I enjoy looking at exhibitions of modern art very much.

55. **Careers for Women** (check one)

_____ I am very much in favor of women pursuing careers.
_____ I am in favor of women pursuing careers.
_____ I am mildly in favor of women pursuing careers.
_____ I am mildly opposed to women pursuing careers.
_____ I am opposed to women pursuing careers.
_____ I am very much opposed to women pursuing careers.

56. **Men's Adjustment to Stress** (check one)

_____ I strongly believe that men adjust to stress better than women.
_____ I believe that men adjust to stress better than women.
_____ I feel that perhaps men adjust to stress better than women.
_____ I feel that perhaps men do not adjust to stress better than women.
_____ I believe that men do not adjust to stress better than women.
_____ I strongly believe that men do not adjust to stress better than women.

INTERPERSONAL JUDGEMENT SCALE

1. Intelligence (check one)

_____ I believe that this person is very much above average in intelligence.
_____ I believe that this person is above average in intelligence.
_____ I believe that this person is slightly above average in intelligence.
_____ I believe that this person is average in intelligence.
_____ I believe that this person is slightly below average in intelligence.
_____ I believe that this person is below average in intelligence.
_____ I believe that this person is very much below average in intelligence.

2. Knowledge of Current Events (check one)

_____ I believe that this person is very much below average in his(her) knowledge of current events.
_____ I believe that this person is below average in his(her) knowledge of current events.
_____ I believe that this person is slightly below average in his(her) knowledge of current events.
_____ I believe that this person is average in his(her) knowledge of current events.
_____ I believe that this person is slightly above average in his(her) knowledge of current events.
_____ I believe that this person is above average in his(her) knowledge of current events.
_____ I believe that this person is very much above average in his(her) knowledge of current events.

3. Morality (check one)

_____ This person impresses me as being extremely moral.
_____ This person impresses me as being moral.
_____ This person impresses me as being moral to a slight degree.
_____ This person impresses me as being neither moral nor particularly immoral.
_____ This person impresses me as being immoral to a slight degree.
_____ This person impresses me as being immoral.
_____ This person impresses me as being extremely immoral.

4. Adjustment (check one)

_____ I believe that this person is extremely maladjusted.
_____ I believe that this person is maladjusted.
_____ I believe that this person is maladjusted to a slight degree.
_____ I believe that this person is neither particularly maladjusted nor particularly well adjusted.
_____ I believe that this person is well adjusted to a slight degree.
_____ I believe that this person is well adjusted.
_____ I believe that this person is extremely well adjusted.

5. Personal Feelings (check one)

_____ I feel that I would probably like this person very much.
_____ I feel that I would probably like this person.
_____ I feel that I would probably like this person to a slight degree.
_____ I feel that I would probably neither particularly like nor particularly dislike this person.
_____ I feel that I would probably dislike this person to a slight degree.
_____ I feel that I would probably dislike this person.
_____ I feel that I would probably dislike this person very much.

6. Working Together in an Experiment (check one)

_____ I believe that I would very much dislike working with this person in an experiment.
_____ I believe that I would dislike working with this person in an experiment.
_____ I believe that I would dislike working with this person in an experiment to a slight degree.
_____ I believe that I would neither particularly dislike nor particularly like working with this person in an experiment.
_____ I believe that I would enjoy working with this person in an experiment to a slight degree.
_____ I believe that I would enjoy working with this person in an experiment.
_____ I believe that I would very much enjoy working with this person in an experiment.

TOPIC 12: QUASI-EXPERIMENTAL DESIGNS

Topic 12 is also concerned with the design of experiments. Here the goal is to provide you with experience in designing quasi-experimental studies that must be conducted within the natural setting of the real world.

In quasi-experimental research the investigator does not have the degree of control over antecedent conditions or extraneous variables that he or she may have within the confines of the laboratory. Whenever we move outside the laboratory we still typically have control over the presentation of the independent variable. However, we frequently loose the ability to randomly assign subjects to experimental treatment conditions. As a result, our subjects are probably not equated at the outset of the experiment and thus a quasi-experiment is the only type of experiment that can be conducted. As far as possible the experimenter should attempt to control for extraneous variables. Since the natural environment has many more extraneous variables potentially affecting the independent variables, the experimenter must always be prepared to evaluate any such variables noted during the course of the experiment. The quasi-experimental design must be constructed in such a manner as to rule out any rival hypothesis whose effect cannot be controlled.

For each of the laboratory exercises in this Topic, you will be given background material on a specific problem area. Your task is to read the introductory material and the statement of the problem and then to design an experiment which would provide an answer to the research problem being asked. Each of these quasi-experiments are to be designed so that they are carried out in a natural setting. Your instructor will assign all or some of the following laboratory exercises.

LABORATORY EXERCISE 1: THE EFFECTS OF ROLE CHANGE ON ATTITUDES

Each individual, engaging in everyday activities, is required to assume at different times a variety of different roles, including spouse, parent, employer or employee, and friend. Each role has an accompanying set of expected behaviors. Role theory makes the assumption that a person's attitudes will be influenced by the role which he or she occupies. If a person made a change in roles from being a renter to a home-owner, there might be a corresponding change in attitude toward care of the home. This change should be one that is congruent with being a home owner.

Although this assumption of role theory is quite logical, there has been little experimental support for its existence. Some correlational evidence does, however, exist. For example, it has been found that commissioned officers are more favorably disposed toward the army than are enlisted men. The problem with such evidence is that it is correlational, and therefore, does not rule out the possibility that the men who are initially proarmy are the ones who become commissioned, rather than being commissioned leading to positive attitudes. Consequently, there is no clear-cut evidence of the causal relationships. In order to obtain such evidence it would be necessary to conduct a longitudinal study where role changes naturally occurred for some individuals. One such setting would be that of an industrial organization that promoted individuals to the level of supervisor from within the ranks of their employees.

Assume that you had identified a company that met such a standard and would allow you to conduct a study which would test this proposition of role theory. Assume also that the company was unionized and that there were approximately as many workers who served as supervisors as served as stewards in the union (about 150 of each). From role theory one would predict that the workers who became stewards would develop more prounion attitudes and the workers who became supervisors would develop a more procompany attitude.

Your task is to design an experiment that would test these predictions which were derived from role theory.

On a separate sheet of paper, write the method section following the APA Publication guidelines. A method section includes the subsections of subjects; apparatus, instruments, or other materials; and procedure. Turn in the completed method section to your instructor.

LABORATORY EXERCISE 2: EMPLOYEE PARTICIPATION IN INCENTIVE PAY PLAN

In the past, many incentive pay plans have been imposed by a variety of companies in an attempt to accomplish the goals of increased productivity and sales, and reduced cost, absenteeism, and turnover. Prior research that has investigated the relative impact of such attempts has produced extremely conflicting results. One study would demonstrate the effectiveness of a given incentive pay plan whereas another study would yield results demonstrating its ineffectiveness. Such conflicting results suggest that other factors are having a moderating effect on the pay incentive plan, and research needs to be conducted to isolate the variables that make a given plan effective for one company but not for another company. One such moderating factor could be the way in which a given pay plan is developed and introduced to the employees. Pay incentive plans could, for example, be developed by the management and imposed upon the employer. On the other hand, pay incentive plans could be developed by the employees and subjected to approval by management. If the employees develop the pay plan they are more likely to be committed to its success as well as having greater understanding of it. Additionally, it is more likely to be appropriate to the employees' working situation. Therefore, the present study will investigate the relative effectiveness of pay plans developed by employees as opposed to those developed by management. It is hypothesized that the most effective pay plan will be the one that the employees develop.

Your task is to design an experiment that would investigate this hypothesis.

On a separate sheet of paper, write the method section following the APA Publication Manual guidelines. A method section includes the subsections of subjects; apparatus, instruments, or other materials; and procedure. Turn in the completed method section to your instructor.

LABORATORY EXERCISE 3: INTELLIGENCE AND ACHIEVEMENT -- WHICH IS THE CAUSAL AGENT?

In looking at the literature on cognitive development one finds two opposing models of mental growth. On the one hand, there is the model that purports that intelligence causes achievement. Such a model was advocated by men such as Francis Galton, Charles Darwin, and Cyril Burt, and supported by the results of numerous studies of twins reared separately. These studies have consistently revealed that the correlation of the intelligence of identical twins reared separately is consistently greater than that of siblings reared apart and also greater than that found between nonrelated persons. If the causal relationship was from achievement to intelligence then the intelligence test scores of these twins should not exceed chance and definitely not exceed that of nonrelated persons or siblings reared apart.

The other model holds that the causal relationship is from achievement to intelligence. Such a model would be emphasized by such individuals as Piaget. Although Piaget recognized the basic importance of inborn processes, he relegated them to a secondary position in the determination of intelligence and achievement. Of primary importance was the acquisition of skills, rules, and information, which when combined resulted in the formation of higher order and more abstract and generalized principles or, in other words, resulted in greater intelligence.

This controversy of the causal order of the intelligence-achievement sequence reveals that the issue has yet to be solved. The basic issue to be solved here is identification of the preponderant causal relationship. It is highly unlikely that one of the two models of mental growth is totally inaccurate and the other totally accurate. Rather, a more probable picture is that intelligence and achievement are both causally related in a feedback loop system. However, one of these two is undoubtedly the more important causal agent.

Your task is to design a study that will investigate the causal relationship between intelligence and achievement to determine which factor is the preponderant causal agent.

On a separate paper of paper, write the method section following the APA Publication Manual guideline. A method section includes the subsections of subjects; apparatus, instruments, or other materials; and procedure. Turn in the completed method section to your instructor.

LABORATORY EXERCISE 4: JAY WALKING BEHAVIOR

Assume that you had observed that some people violate traffic signals and cross the street in spite of the fact that the traffic signal is flashing "wait". You want to know what factors motivate individuals to violate such prohibitions. However, in designing such a study it would be very difficult to randomly assign subjects to appear at various street crossings at different times during the day to correspond with your experimental treatment conditions. Therefore, the design in most instances must be of the quasi-experimental variety.

Lefkowitz, Blake, and Mouton (1955) used such a design several decades ago. They postulated that one of the variables motivating jaywalking behavior was the presence of a model. In other words, if another person jaywalked first you may be also more prone to jaywalk. They found, in investigating this phenomenon, that the presence of such a model is effective only when the model is a high status model. Mullen, Copper and Driskell (1990) reviewed seven studies that examined the effects of model behavior on pedestrian jaywalking.

What other factors might influence a person's decision to jaywalk? Your task is to identify at least three possible variables which you hypothesize have an influence upon jaywalking. Then design an experiment which would allow you to test experimentally the influence of these variables. In designing the experiment you must decide on a number of factors, such as the following: (1) At what point has a person jaywalked? Is it when they have completely crossed the street? What if the person only crossed the street part of the way and then turned around and returned to the curb? (2) What individuals will you include as potential subjects for the experiment? People may be continuously approaching the curb. Be attentive to possible confounds.

Listed below is some background reading if you wish to pursue this topic more thoroughly. This experiment could be carried out if you informed the local police department of your intent and obtained their permission, as well as the permission of your human subjects' committee.

Jaywalking References:

Jason, L. A., & Liotta, R. F. (1982). Pedestrian jaywalking under facilitating and nonfacilitating conditions. *Journal of Applied Behavior Analysis, 15*, 469-473.

Jorgensen, N. D. (1988). Risky behavior at traffic signals: A traffic engineer's view. Special issue: Risky decision-making in transport operations. *Ergonomics, 31*, 657-661.

Lefkowitz, M., Blake, R. R., & Mouton, J. S. (1955). Status factors in pedestrian violation of traffic signals. *Journal of Abnormal and Social Psychology, 51*, 704-706.

Mullen, B., Copper, C., & Driskell, J. E. (1990). Jaywalking as a function of model behavior. *Personality and Social Psychology Bulletin, 16*, 320-330.

Russell, J. C., Wilson, D. O., & Jenkins, J. F. (1976). Informational properties of jaywalking: An extension to model sex, race and number. *Sociometry, 39*, 270-273.

Segelman, C. K., & Segelman, L. (1976). Authority and conformity: Violation of a traffic regulation. *The Journal of Social Psychology, 100*, 35-43.

van Houton, R. (1988). The effects of advance stop lines and sign prompts on pedestrian safety in a crosswalk on a multilane highway. *Journal of Applied Behavior Analysis, 21*, 245-251.

TOPIC 13: SINGLE-SUBJECT AND SINGLE-GROUP DESIGNS

The laboratory goals of Topic 13 are to provide you with experience in identifying and designing single-subject and single-group designs.

As noted in the preface to *Methodological and conceptual issues in applied behavior analysis: 1968-1988* (Iwata et al., 1989),

> Originally developed in operant conditioning laboratories, the single-subject approach represents a clear departure from research methods typical of most other areas of psychology and the social sciences, by placing greater emphasis on the objective measurement of ongoing behavior, the collection of extended data samples per individual subject, the use of a subject's behavioral "baseline" as a control condition for the purpose of conducting experimental comparisons, and the evaluation of experimental effects through visual examination of data (p.iii).

Single-subject and single-group studies use some form of a time-series design. Repeated measurements are taken on the dependent variable both before and after a treatment condition is introduced. The most basic design is an <u>A-B-A design</u> with three conditions: A - the baseline; B - the experimental treatment; and, A - the second baseline that is reintroduced without the treatment. For further information about this basic A-B-A design, refer to your textbook for discussion and examples.

Another approach is the <u>multiple-baseline design</u>. Often in an A-B-A design the second baseline condition does not return to the original baseline. This creates ambiguous results so that one cannot completely eliminate a rival hypothesis. To overcome this deficit, some researchers use a multiple baseline design and examine the same person (or group) over several situations, different people (groups) over the same situations, or several behaviors for the same individual. This design might involve taking a baseline measure on two subjects and then administering the treatment condition to one of the subjects while the other remained in the baseline condition. Only after the treatment condition had been administered to the first subject would it be administered to the second subject.

More elaborate designs are also found in the literature. <u>Interaction designs</u> permit the experimenters to evaluate different levels of the independent variables (treatment) in comparison to the baseline condition. For instance, a researcher might examine the effect of verbal reinforcement ("very good") and food reinforcement (candy) on eye contact with autistic children. The variables could be examined alone and with one another, in comparison to the baseline phase(s).

Finally, another design is the <u>changing-criterion design</u>. The target behavior is examined first in a baseline condition. Then a treatment is applied until a certain criterion is obtained. Once the initial criterion of successful performance is met, the experimenter changes the criterion level until the altered criterion is met. For instance, a disruptive child in the classroom might be given a reward (playing games on a computer) if he or she sits still 15% of the time. Once that criterion is met, the teacher may change the criterion to 25% of the time, and so on.

LABORATORY EXERCISE 1: EVALUATING SINGLE-SUBJECT DESIGNS IN PUBLISHED RESEARCH

The purpose of this laboratory exercise is to give you experience in identifying and analyzing single subject designs.

Your assignment for this laboratory exercise is to go to the library and find a single-subject design experiment. Excellent journals which often report such studies are *Behavior Therapy, Experimental Analysis of Behavior*, and *Journal of Applied Behavior Analysis*.

Answer the following on a separate sheet of paper and hand into your instructor.

1. Give the complete reference of the study.

2. What was the purpose of the study?

3. What were the hypotheses?

4. Describe the method in detail.

    a. Who was (were) the subject(s)? How was the subject chosen?

    b. Identify which kind of design was used. Describe it in detail.

    c. Identify the independent variable(s). If it changes in any manner during the course of the experiment, explain how it did.

    d. Identify the dependent variable(s).

5. Describe the major results.

6. Describe the conclusions of the author(s). Are they warranted?

7. Are there any rival interpretations of the results? If so, discuss them briefly.

LABORATORY EXERCISE 2: DESIGNING SINGLE-SUBJECT STUDY -- INCREASING SAFETY SEAT BELT USAGE

The purpose of this laboratory exercise is to give you experience in evaluating a single-group study and in designing a subsequent study.

Behavior change interventions are used not only in clinical settings but also in public domains by community and applied experimental psychologists. Examples of such intervention research programs include increasing the collection of recyclables (e.g., Geller, Winett, & Everett, 1982), increasing conservation behaviors (e.g., Dussault, 1990; Geller et al., 1982), increasing safety belt usage (e.g., Geller, 1988) reducing cigarette smoking (Schinke, Gilchrist, & Snow, 1985), reducing AIDS risk behaviors (Kelly, St. Lawrence, Hood, & Brasfield, 1990), and reducing pregnancy among teenagers (Schinke, Blythe, Gilchrist, & Burt, 1981).

A major public health problem is automobile crash-induced injuries, many of which could be reduced in severity if automobile occupants had used safety belts. Legislature has been enacted in numerous states, using negative reinforcement and punishment (punative fines), to insure that automobile occupants use safety belts and still people often do not follow the law. As Thyer and Geller (1987) note, "Such belt use mandates are usually only moderately effective, unless they are rigorously enforced (see Jonah et al., 1982), and they have the disadvantage of imposing additional elements of aversive control on citizens already burdened with a surfeit of punitive legal sanctions" (p. 485). Rather, they advocate the use of positive reinforcement contingencies to get the general public to comply with buckling up.

Psychologists have used a number of techniques to motivate safety belt usage with success (Streff & Geller, 1986). Geller (1988) refers to an "ABC" model for behavior change: activator -- behavior -- consequence. Activators (reminder, prompt, model education, commitment, incentive, disincentive) produce a behavior (safety belt use) which in turn results in a consequence (reward or positive reinforcer, punisher or negative reinforcer). The use of a vehicle dashboard sticker (a prompt) that read "SAFETY BELT USE REQUIRED IN THIS VEHICLE" has substantially increased safety belt use in a study by Thyer and Geller (1987). They used an A (baseline) - B (intervention of stickers) - A (withdrawal of stickers) - B (intervention of stickers) design. Safety belt use ranged from 17% to 50% (Mean = 34%) of the passengers in the first baseline. By the end of the second intervention phase, safety belt use ranged from 54% to 90% (Mean = 78%) of the front-seat passengers. The figure[1] at the end of this section shows the percentage of vehicle passengers who bucked up in the different conditions.

On the following pages is reprinted the complete article by Thyer, Geller, Williams, and Purcell (1987) in which they used an A-B-A-B design to study the effectiveness of a community-based prompting intervention on a university campus. You are to read critically this article and then design a similar single-group A-B-A-B study that would be feasible to carry out on your campus or in your community. Pay close attention to the design of the prompts and how you will make them visible to drivers. Consider issues for future research that Thyer et al. present in their discussion section, and determine how you can address them in your study. Write up the method section for this proposed study.

---

[1]From Thyer and Geller (1987, p. 490). Copyright 1987 by Sage Publication, Inc. Reprinted by permission.

We hope that you will actually carry out the study in such a manner that it is publishable in a psychological journal. From research laboratory courses such as the one you are presently in come many projects that can be carried out in future semesters as independent research projects. Large and colorful "Flash-for-Life" cards are available from E. Scott Geller (Department of Psychology, Virginia Polytechnic Institute and State University, Blacksburg, VA 24061) for a nominal cost.

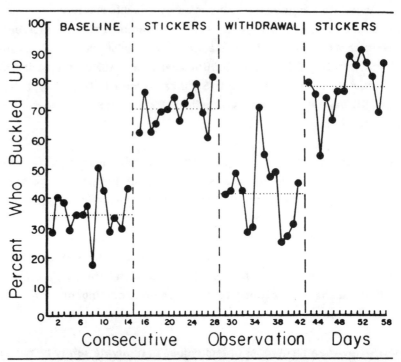

The percentage of passengers who buckled up over 58 consecutive observation days, two-weeks per consecutive baseline, intervention, withdrawal, and intervention phase.

References:

Dussault, C. (1990). Effectiveness of a Selective Traffic Enforcement Program combined with incentives for seat belt use in Quebec. Special Issue: Traffic injury prevention. *Health Education Research, 5*, 217-223.

Geller, E.S. (1988). A behavioral science approach to transportation safety. *Bulletin of the New York Academy of Medicine, 64*, 632-661.

Geller, E.S., Winett, R.A., & Everett, P.B. (1982). *Preserving the environment: New strategies for behavior change.* New York: Pergamon.

Johan, B.A., Dawson, N.E., & Smith, G.A. (1982). Effects of a selective traffic enforcement program on seat belt use. *Journal of Applied Psychology, 67*, 84-96.

Kelly, J.A., St. Lawrence, J.S., Hood, H.V., & Brasfield, T.L. (1990). Behavioral interventions to reduce AIDS risk activities. *Journal of Consulting and Clinical Psychology, 57*, 60-67.

Schinke, S.P., Blythe, B.J., Gilchrist, L.D., & Burt, G.A. (1981). Primary prevention of adolsecent pregnancy. *Social Work with Groups, 4*, 121-135.

Schinke, S.P., Gilchrist, L.D., & Snow, W.H. (1985). Skills intervention to prevent cigarette smoking among adolescents. *American Journal of Public Health, 75*, 665-667.

Streff, F.M., & Geller, E.S. (1986). Strategies for motivating safety belt use: The application of applied behavior analysis. *Health Education Research: Theory and Practice, 1*, 47-59.

Thyer, B.A., & Geller, E.S. (1987). The "buckle-up" dashboard sticker: An effective environmental intervention for safety belt promotion. *Environment and Behavior, 19*, 484-494.

Thyer, B.A., Geller, E.S., Williams, M., & Purcell, E. (1987). Community-based "flashing" to increase safety belt use. *The Journal of Experimental Education, 55*, 155-159.

Community-Based "Flashing" to Increase Safety Belt Use

BRUCE A. THYER
Florida State University

E. SCOTT GELLER
Virginia Polytechnice Institute
and State University

MELVIN WILLIAMS
ELAINE PURCELL
Florida State University

ABSTRACT

A community-based prompting intervention for safety belt promotion was field tested at two parking lots on a large university campus. The intervention involved a co-ed displaying a flash card that read, "Please buckle up —I care" to unbuckled drivers of vehicles exiting the parking lots. If the driver buckled up, the "flasher" flipped the card over and displayed the message, "Thank you for buckling up." Drivers who were already wearing a shoulder belt when exiting the parking lot were shown only the "thank you" side of the flash card. Simultaneous ABAB paradigms over a 4-week period demonstrated functional control of safety belt use at each parking lot by this prompting intervention. A total of 1,260 flashing episodes occurred and overall compliance with the buckle-up request was 25%. Important issues for follow-up research are discussed, especially the need for further study of the post-intervention, residual effect of prompting that was observed.

EACH YEAR in the United States motor vehicle accidents result in at least 45,000 deaths and 500,000 serious injuries (Bigelow, 1982). In fact, vehicle crashes are the leading cause of fatalities among Americans aged 5 to 34 (Sleet, 1984), and the financial liability of U.S. traffic accidents exceeds $60 billion per year (Pabon, Sims, Smith, & Associates, 1983). It is estimated that 55% of all traffic fatalities and 65% of all injuries would be prevented if vehicle safety belts were used (Department of Transportation, 1983); yet the majority of the Americans do not use this proven safety device.

As a result of the current nationwide efforts to promote safety belt use in the U.S., belt wearing is gradually increasing. Systematic and comprehensive observations of safety belt use by the National Highway Traffic Safety Administration revealed 10.9% belt wearing from 1977 to 1979 (Two Year Study, 1980), and 13.6% safety belt use in 1983 (Steed, 1983). Recently, safety belt use has reached 50% in states that have a safety belt use law (Insurance Institute for Highway Safety, 1985), and has exceeded 60% among employees of corporations that have implemented safety belt incentive programs (e.g., see reviews by Geller, 1984, 1985).

As reviewed recently by Streff and Geller (1986), various strategies have been employed to motivate safety belt use in the U.S., including (a) engineering approaches (e.g., buzzer/light reminders and ignition interlock systems that prevent vehicles from starting unless front seat belts are fastened); (b) legal mandates (from belt use policies at industries and institutions to statewide belt use laws); (c) mass media campaigns (from community billboards to TV and radio spots); (d) education and awareness sessions at industrial sites, schools, civic organizations, and churches; (e) incentive/reward programs at specific corporate, business, and government locations and throughout entire communities; and (f) small-scale reminder strategies (i.e., including highway signs, bumper and dashboard stickers, flyers, and "mailouts").

The present research evaluated the impact of a particular reminder strategy that extended a technique developed recently by Geller and his students, entitled "Flash-for-Life." As described by Geller (1985), a passenger (termed a "flasher") in a stopped vehicle holds a large flash card so that an unbuckled driver or passenger of another stopped vehicle can see the printed message "Please buckle up—I care." If the unbuckled occupant sees the message and buckles up, the flasher flips the card over to reveal the bold message "Thank you for buckling up." In one field study, this simple reminding technique was successful in getting 22% of 893 vehicle occupants who looked at the flash card to buckle up on the spot (Geller, Bruff, & Nimmer, 1985). In the present study, the flash card was not displayed from a vehicle but rather was shown to vehicle occupants as they exited parking lots. This simple modification of the Flash-for-Life technique more effectively reached large numbers of drivers and passengers in a short period of time.

Method

Participants and Setting

The study was conducted on the campus of Florida State University (FSU) during the fall of 1985. Located in Tallahassee, Florida (population 170,000), FSU has approximately 22,000 students and employs 1,700 faculty and staff. At the time of this study the state of Florida did not have a mandatory safety belt use law for adults. Two faculty and staff parking lots adjacent to the school of social work were chosen as intervention sites. The Call Street lot had a capacity of 115 automobiles and could only be entered if the driver had a plastic key-card to open the entrance gate. Key-cards were issued to all faculty and staff who purchased a university parking decal. The Dogwood Way lot had a capacity of 75 automobiles and, although posted signs restricted its use to faculty and staff with parking decals on their automobiles, it could be entered by any driver without such a permit who was willing to risk a $10.00 fine. Both lots were routinely filled to capacity each weekday.

Data Collection

Two graduate students stationed across the street from the single exit of each lot independently recorded whether or not the driver of an exiting vehicle was wearing a shoulder belt. These observations took place Monday through Friday for 4 consecutive weeks from 4:00 P.M. to 5:00 P.M. (the daily peak period of exiting vehicles). Interobserver agreement was calculated using the following formula:

$$A = [agreements/(agreements + disagreements)] \times 100$$

The observers were trained to at least an 80% agreement rating prior to collecting data. Each day the observer-pairs randomly assigned one member to be the primary observer and the other to be the reliability observer.

Intervention

Each parking lot had a permanent stop sign on the passenger's side of exit. As each exiting automobile approached the exit, a female graduate student (i.e., the "flasher"), standing on the driver's side of the parking lot exit, displayed an 11 × 14 inch Flash-for-Life card with the printed message "Please buckle up—I care" to the oncoming driver. The sign was held chest high. If the driver was already wearing a shoulder belt, or was observed buckling up, the flasher reversed the sign to display the message "Thank you for buckling up." Figure 1 illustrates the front and back of the flash card. Flashers did not attempt any other method to prompt or reward safety belt use (e.g., verbal pleas to buckle up or shouts of "thank you").

Design

Baseline. During the first week of data collection, observer-pairs unobtrusively recorded the number of

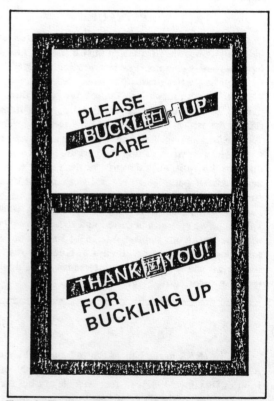

Figure 1—The front and back of the Flash-for-Life card. On both sides the background is white with a 1.3 cm, yellow border; the safety belt is black with yellow letters (3 cm high); the rest of the letters are black (also 3 cm high).

vehicles exiting each lot and the number of drivers wearing a shoulder belt. Flashers were not present.

Intervention. During the second week, a flasher displayed a Flash-for-Life card as described previously, and safety belt use was recorded after the vehicle passed the flasher.

The baseline and intervention phases were repeated during the third and fourth weeks, respectively. Thus, the experimental design for each lot represented an ABAB time series (Barlow & Hersen, 1984).

Results

Interobserver Agreement

The number of vehicles exiting daily from the Dogwood Way lot ranged from 49 to 78, with a mean of 63 and totaling 1,250 during 4 weeks. The daily number of exiting vehicles from the Call Street lot ranged from 43 to 78 and also had a mean of 63 exits per day, with a total of 1,268 over 4 weeks. A reliability observer was present for all of the observation periods at the Dogwood Way lot, and for 95% of the observation sessions at the Call Street lot. Daily interobserver reliability ranged from 86% to 100% agreement ($M = 96\%$) at the Dogwood Way lot and from 89% to 100% ($M = 95\%$) at the Call Street lot.

Shoulder Belt Use

Figure 2 shows the daily percentage of shoulder belt use by drivers exiting the two parking lots observed during each of the four experimental conditions. For the Call Street lot, daily shoulder belt use ranged from 19% to 26% ($M = 22\%$) during the first baseline; from 45% to 65% ($M = 52\%$) during the first intervention week; from 34% to 41% ($M = 37\%$) during the second baseline; and from 54% to 68% ($M = 61\%$) during the second intervention.

For vehicles exiting the Dogwood Way lot, driver shoulder belt use ranged from 10% to 21% ($M = 17\%$) during the first baseline; from 32% to 47% ($M = 39\%$) during the first intervention week; from 14% to 25% ($M = 20\%$) during the second baseline week; and from 33% to 52% ($M = 42\%$) during the second intervention phase.

The results demonstrate a relatively powerful effect for the simple prompting intervention. Although the overall number of percentage points increased by the flashing intervention was slightly greater at the Call Street lot than the Dogwood Way lot (i.e., mean increase of 27% vs. 22%, respectively), the overall increase from baseline to intervention was actually higher at the Dogwood Way lot (i.e., 91.5% at Call Street vs. a 119% increase in belt wearing at Dogwood Way). In other words, the prompting procedure almost doubled shoulder belt use by drivers exiting the Call Street lot

and more than doubled the rate of belt use by drivers exiting Dogwood Way.

Functional control of the intervention is clearly shown at both lots by the dramatic decrease in shoulder belt use when prompting was withdrawn (during the second baseline phase) and by the immediate increase in belt wearing at the beginning of the second intervention phase. It is noteworthy, however, that the reversal was not complete, implying that some drivers continued to buckle up after the prompt was removed. The residual effect was markedly higher at the Call Street lot than the Dogwood Way lot. Specifically, at Call Street the second baseline was 15 percentage points higher than the initial baseline (a 68% increase in belt use); whereas at Dogwood Way, the second baseline was only 3 percentage points higher than the initial baseline (a 17.6% increase over baseline).

Social Validity

The flashers reported that occasionally drivers of exiting vehicles paused to offer comments or ask questions. Virtually all such comments were of an approving nature (e.g., thanking the flasher for the reminder to buckle up). Some drivers questioned why the co-ed was engaged in the prompting activity, to which a standard reply was given (i.e., "This is for a class project in the School of Social Work"). Only one driver paused to offer a negative comment, indicating disapproval of safety belts because an acquaintance was trapped inside a flaming vehicle following a collision.

A local television station prepared an on-site news segment about the project, which was aired on three occasions after the study was completed. A local radio station broadcast a similar interview with the senior author, describing the study and the importance of promoting consistent safety belt use.

Discussion

While developing the Flash-for-Life technique, Geller et al. (1985) considered certain characteristics of a verbal or written message that successfully prompted behaviors related to environmental preservation (see review by Geller, Winett, & Everett, 1982). Specifically, the prompt (or message) should be polite, refer to a specific behavior, and occur in close proximity to the requested behavior. Also, the requested behavior should be relatively convenient to emit. In the original Flash-for-Life intervention, Geller et al. (1985) attributed part of their prompting success to the involvement of modeling. That is, the flasher was wearing a readily visible shoulder belt.

The present study showed more pronounced intervention effects than Geller et al. (1985) with an application of the Flash-for-Life technique that did not include modeling and was more cost effective in reaching large

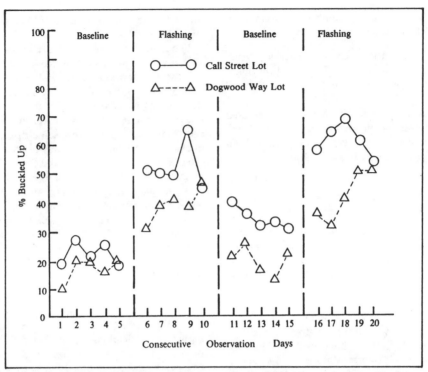

Figure 2—Percentage of vehicle drivers wearing a shoulder belt while exiting two campus parking lots during two baseline periods and two intervention phases when a co-ed displayed a Flash-for-Life card.

numbers of vehicle occupants. In particular, the vehicle-based flashing by Geller et al. took approximately 100 hours to prompt 1,087 unbuckled drivers, and of this total, 17.7% looked at the flash card and complied with the buckle-up request. In the present study, 1,260 drivers were "flashed" in only 20 hours, and although some of these drivers were already buckled up, comparisons with baseline observations indicate that the compliance rates were actually larger than those found by Geller et al. In fact, the overall estimated compliance rate in the present study was 25%. This intervention impact is larger than that found by Geller et al. and is comparable to the outcomes of prior prompting studies that increased the purchase of returnable soft drink containers (Geller, Farris, & Post, 1973) and the appropriate disposal of litter (Geller, Witmer, & Orebaugh, 1976) with specific and polite messages, delivered at the time and place when the requested behavior was convenient to emit.

It is encouraging that shoulder belt use during the second baseline phase was higher than initial baseline levels, especially at the Call Street lot. This suggests that the prompting intervention had more than transient impact on some drivers. The greater residual effect at Call Street than Dogwood Way (i.e., 68% vs. 17.6% in-

creases over initial baseline, respectively) may have been partially due to the more homogeneous and consistent sample of drivers (i.e., only faculty and staff) exiting the electronically gated Call Street lot. Although systematic records were not maintained, observers did note a substantial number of non-decaled and student-decaled vehicles exiting the Dogwood Way lot. Previous safety belt research has demonstrated significant response maintenance advantages when individuals received repeated rewards for being buckled up (Geller, 1983); and other campus-based programs have found faculty and staff to be much more responsive than students to incentives for safety belt use (Geller, Kalsher, Lehman, & Rudd, 1986; Rudd & Geller, 1985).

Rudd and Geller (1985) implemented a campuswide incentive program for safety belt promotion during a consecutive 3-week period within each quarter of the 1983–84 academic year. Each withdrawal phase at the end of the fall, winter, and spring quarters showed increasing residual effects. In other words, when each incentive program was withdrawn, campuswide safety belt use decreased prominently but remained significantly above the pre-intervention baseline of that quarter. And, this stepwise residual effect was greater for faculty and staff than students. Geller et al. (1986)

138

replicated these findings during the 1984–85 academic year with a different incentive approach toward safety belt promotion.

The present study of campus-based prompting showed stepwise residual effects similar to those observed with the application of incentives. However, it is noteworthy that the phase durations were quite short and there was only one withdrawal phase per parking lot. It would certainly be instructive to alternate several baseline and flashing phases over an extended time period. Furthermore, the recording of vehicle license plate numbers would enable a precise evaluation of repeated flashes per individual driver. Future research should also address the necessary components of the flashing intervention. For example, it is likely that the active display and manipulation of the flash card by a flasher was necessary, because in two previous studies, patrons leaving a bank exchange window did not buckle up when handed a flyer that politely requested safety belt use (Geller, Johnson, & Pelton, 1982; Johnson & Geller, 1984).

In summary, the present study demonstrated substantial behavior change potential of a simple prompting technique that could be integrated readily with large-scale education, legal, and incentive approaches to safety belt promotion, and thereby increase the impact of a comprehensive safety belt program. Also, the research raised important empirical questions regarding the long-term, repeated impact of the Flash-for-Life technique. The need for follow-up study is underlined by the simplicity and low cost of this prompting strategy and by the life-saving potential of an intervention that increases the wearing of vehicle safety belts.

NOTES

The authors are grateful for the helpful comments and suggestions by Galen R. Lehman on an earlier draft of this paper.

Requests for reprints should be sent to E. Scott Geller, Department of Psychology, Virginia Tech, Blacksburg, VA 24061.

Requests for "Flash-for-Life" cards for research purposes should be sent to E. Scott Geller at the above address.

REFERENCES

Barlow, D. H., & Hersen, M. (1984). *Single case experimental designs: Strategies for studying behavior change* (2nd ed.). Elmsford, NY: Pergamon Press.

Bigelow, B. E. (1982). The NHTSA Program of Safety Belt Research. *SAE Techincal Paper Series*, No. 820797. Warrendale, PA: Society of Automotive Engineers.

Department of Transportation. (1983, October). Federal motor vehicle safety standards: Occupant crash protection. *Federal Register*, *48*(203).

Geller, E. S. (1983). Rewarding safety belt usage at an industrial setting: Tests of treatment generality and response maintenance. *Journal of Applied Behavior Analysis, 16*, 43–56.

Geller, E. S. (1984). Motivating safety belt use with incentives: A critical review of the past and a look to the future. *SAE Technical Paper Series*, No. 840326. Warrendale, PA: Society of Automotive Engineers.

Geller, E. S. (1985). *Corporate safety belt programs.* Blacksburg, VA: Virginia Polytechnic Institute and State University.

Geller, E. S., Bruff, C. D., & Nimmer, J. G. (1985). "Flash for life": Community-based prompting for safety belt promotion. *Journal of Applied Behavior Analysis, 18*, 309–314.

Geller, E. S., Farris, J. C., & Post, D. S. (1973). Prompting a consumer behavior for pollution control. *Journal of Applied Behavior Analysis, 6*, 367–376.

Geller, E. S., Johnson, R. P., & Pelton, S. L. (1982). Community-based interventions for encouraging safety belt use. *American Journal of Community Psychology, 10*, 183–195.

Geller, E. S., Kalsher, M. J., Lehman, G. R., & Rudd, J. R. (1986). *A universitywide safety belt program: Using rewards to motivate a buckle-up commitment.* Manuscript in preparation. Blacksburg, VA: Virginia Polytechnic Institute and State University.

Geller, E. S., Winett, R. A., & Everett, P. B. (1982). *Preserving the environment: New strategies for behavior change.* Elmsford, NY: Pergamon Press.

Geller, E. S., Witmer, J. F., & Orebaugh, A. L. (1976). Instructions as a determinant of paper-disposal behaviors. *Environment and Behavior, 8*, 417–438.

Insurance Institute for Highway Safety. (1985, November). *The highway loss reduction status report, 20*(12).

Johnson, R. P., & Geller, E. S. (1984). Contingent versus noncontingent rewards for promoting seat belt use. *Journal of Community Psychology, 12*, 113–122.

Pabon, Sims, Smith, & Associates, Inc. (1983). *Motivation of employers to encourage their employees to use safety belts: Phase II* (Contract No. DTNH 22-80-C-07439). Washington, DC: National Highway Traffic Safety Administration.

Rudd, J. R., & Geller, E. S. (1985). A university-based incentive program to increase safety belt use: Toward cost-effective institutionalization. *Journal of Applied Behavior Analysis, 18*, 215–226.

Sleet, D. A. (1984). A preventative health orientation in safety belt and child safety seat use. *SAE Technical Paper Series*, No. 840325. Warrendale, PA: Society of Automotive Engineers.

Steed, D. K. (1983, September). *Alcohol and safety belts.* Keynote address to the National Conference on Alcohol Countermeasures and Occupant Protection, Denver, CO.

Streff, F. M., & Geller, E. S. (1986). Strategies for motivating safety belt use: The application of applied behavior analysis. *Health Education Research: Theory and Practice, 1*, 47–59.

Two year study shows decrease in safety belt use. (1980). *Emphasis, 10*(3), 4.

LABORATORY EXERCISE 3: DESIGNING SINGLE-SUBJECT -- BEHAVIOR MODIFICATION OF NOCTURNAL ENURESIS (BEDWETTING)

The purpose of this laboratory exercise is to give you experience in designing a single-subject experiment, using one of the designs discussed at the beginning of this Topic.

Enuresis is a common problem that affects 22% of 5 year olds and 10% of 10 year olds (Scipien, Barnhart, Chard, Howe, & Phillips, 1986). Treatment approaches have included drugs or behavior modification procedures. The use of behavior modification procedures to treat nocturnal enuresis (bedwetting) has typically involved use of conditioning methods which include the use of a pad and bell, or pad and shock. When using these procedures the onset of nocturnal micturition is followed by the presentation of shock or the sounding of the bell, which awakens the child. While this is the typical behavioral procedure used, a variety of additional techniques, such as reinforcement of dry nights, have been investigated.

Although the enuretic is most typically associated with nocturnal micturition it has also been demonstrated that he or she also has little bladder control diurnally. Enuretics urinate frequently during the day and cystometrographic records reveal that these individuals have a low bladder capacity with strong urges to urinate at low bladder pressures and volume. Such evidence has suggested to a number of researchers that enuretics do not receive sufficiently strong stimulation from the bladder to awaken the individual during sleep. This deficit has been attributed to a variety of factors, which include inadequate physiological or neural maturations and inadequate learning of micturition patterns.

Regardless of the cause of the low bladder capacity of the enuretics, it has been recommended by a number of researchers that treatment should consist of a program that teaches the enuretic to inhibit urination during the day with the expectation that the effect will generalize to nighttime. Treatment should consist of providing the enuretic with an abundance of fluids and instructing him or her to retain the urine for as long as possible. Such retention is expected to expand the capacity of the bladder to enable one to hold the 10 to 12 oz. of urine required for the output of night urine.

Your task is to design an experiment which will test the effectiveness of this method of retention training conditioning treating enuretics.

Write up this laboratory exercise as a methods section, according to APA Publication Manual guidelines.

References:

Kaplan, S. L., Breit, M., Gauthier, B., & Busner, J. (1989). A comparison of three nocturnal enuresis treatment methods. *Journal of the American Academy of Child and Adolescent Psychiatry, 28,* 282-286.

Ronen, T., Wozner, U., & Rahaw, G. (1992). Cognitive intervention in enuresis. *Child and Family Behavior Therapy, 14,* 1-14.

Scott, M. A., Barclay, D. R., & Houts, A. C. (1992). Childhood enuresis: Etiology, assessment, and current behavioral treatment. *Progress in Behavior Modification, 28.* 83-117.

LABORATORY EXERCISE 4: DESIGNING SINGLE-SUBJECT STUDY -- BEHAVIOR MODIFICATION OF BIZARRE VERBALIZATIONS

The purpose of this laboratory exercise is to give you experience in designing a single-subject experiment, using one of the designs discussed at the beginning of this topic.

In the past it has been recognized that a camp environment may provide an excellent situation for applying behavioral procedures for the remediation of problematic behaviors. Although little research has been conducted in such settings they are considered advantageous from the standpoint of providing a natural setting for conducting controlled research. Additionally, camp environments have the physical facilities and the social environment capable of providing strong reinforcers to children, and the environment is rather novel, which may be conducive to behavior change.

It has just been pointed out that there are many characteristics of a camp setting that make it an excellent environment to administer a behavioral program that remedies problematic behaviors. Assume that, because of lack of systematic evidence to support the above contentions, you wanted to demonstrate the utility of camp settings. Assume further that you had contacted a camp that would allow you to conduct a research study and that they had admitted an 8-year-old minimally brain damaged boy who demonstrated a high frequency of bizarre verbalizations primarily about penguins. Since these verbalizations interfered with his development of good interpersonal relations with adults or peers, you consider him to be an excellent candidate for your study. You decided to use extinction in an attempt to eliminate these bizarre verbalizations in four camp settings: walking on the trail/evening activity, dining hall, cabin, and education. Your task is to design a study that would test the effectiveness of extinction in eliminating these bizarre verbalizations.

Your task is to design a single-subject experiment.

Write up this laboratory exercise as a method section, according to APA Publication Manual guidelines.

LABORATORY EXERCISE 5: DESIGNING A SINGLE-GROUP STUDY -- REDUCING LITTERING BEHAVIOR

The purpose of this laboratory exercise is to give you experience in designing an environmental psychology study which uses an A-B-A design.

Environmental psychologists have become interested in determining ways in which we can change behavior of individuals so that they preserve their environments. An excellent book that summarizes many studies in this area was written by Geller, Winett, and Everett (1982).

As you have probably noticed, littering is a problem along some of our highways. You are assigned to develop a project to evaluate a way to make people stop, or at least, decrease their littering behavior. You can either threaten (possible punishment) or reward (positive reinforcement) non-littering behavior through the use of signs, fines, or other methods.

Your task is to design an A-B-A quasi-experimental study which will evaluate the effectiveness of some treatment on littering behavior. Determine where, when, and under what conditions you will conduct your study. What is (are) your independent variable(s)? What is (are) your dependent variables (s)? How will you operationally define "littering"? What is your procedure?

Write up this laboratory exercise as a method section, according to APA Publication Manual guidelines.

References:

Dixon, R. S., & Moore, D. W. (1992). The effects of posted feedback on littering: Another look. *Behaviour Change, 9*, 83-86.

Geller, W. S., Winett, R. A., & Everett, P. B. (1982). *Preserving the environment: New strategies for behavior change.* New York: Pergamon.

Levitt, L., & Leventhal, G. (1986). Litter reduction: How effective is the New York State Bottle Bill? *Environment and Behavior, 18*, 467-479.

TOPIC 14: ETHICS

The goal for Topic 14 is to introduce you to some of the ethical issues surrounding research by presenting you with a series of experiments to evaluate and by having you evaluate and develop informed consent forms and debriefing forms.

More than two decades ago a great deal of concern was beginning to be expressed over the ethics of human research. This concern developed when a number of investigators became aware of the fact that scientists did not always operate in such a fashion that would benefit others and that experiments were not always conducted in a manner that insured the safety of others. Within the field of psychology the ethical outcry was primarily focused on the deception that was included in many experiments primarily within the field of social psychology. However, within the field of medicine a similar ethical concern was being expressed and this concern was dramatized and epitomized with the publication, in 1972, of a study conducted by the United States Public Health Service over the prior 40 years (Jones, 1981). This study consisted of an investigation of the long term effects of untreated syphilis in a group of black men. The unethical nature of this study can be seen in a variety of instances. The men were not told of the purpose of the study or that they were not being treated for syphilis. In fact, other physicians in the area were contacted and asked not to treat these men if they appeared at their offices requesting treatment.

The result of such concerns and unethical studies was the formulation, by the American Psychological Association and other scientific organizations, of sets of ethical principles to guide the researcher in the conduct of experimentation. Similarly, the government has established a set of guidelines that must be followed in the conduct of any federally funded study. Additionally, each university is required to establish an Institutional Review Board which has as its purpose reviewing all human research to insure that the subjects are treated ethically.

At the present time a similar situation is taking place in the area of animal experimentation, except that the concern over the ethical treatment of animals is coming not only from certain scientists but also from groups such as the Animal Rights Coalition, the Mobilization for Animals, and the Society Against Vivisection. These groups are claiming that animals used in psychological and medical research are given intense and repeated pain, that they die from hunger and thirst, or that their bones are crushed and internal organs are ruptured in experiments. At certain laboratories and at the American Psychological Association annual meetings as well as other scientific meetings, there have been demonstrations against the use of animals in psychological research. As a result, scientists are exploring, in greater depth, the ethical issues relating to such research.

The Rockerfeller University psychologist Neal Miller, among other psychologists, has been a strong advocate of animal research. At the 1984 American Psychological Association annual meeting he pointed out the number of ways behavioral research with animals has benefitted humans. For instance, they help us learn how drugs can treat hypertension and psychoses and how to rehabilitate nerve-damaged limbs. Coile and Miller (1984) surveyed all animal studies published in American Psychological Association journals over a five year period and could not find any studies that involved inhumane treatment of animals. Ten percent of the 608 evaluated studies used electric shock, mostly mild. This does not mean that cruel and inhumane treatment has not existed in psychological research studies using animals. Rather, it suggests that such abuses are infrequent and are the exception and not the rule.

There are a variety of mechanisms which have been set up to insure that research animals are treated in a humane manner. For example, the Federal Animal Welfare Act of 1966 requires the periodic inspection of all animal research facilities. In Great Britain, researchers must obtain licenses in order to carry out animal research. The American Psychological Association's Committee on Animal Research and Experimentation is continuing to evaluate and introduce guidelines that will protect research animals. Thus, the animal rights groups are having an effect on the research community. Articles are being written for the general public (e.g., Hubbell, 1990). The end result will be even more rigid assurance of the humane treatment of research animals. However, it seems unlikely that it will result in the elimination of the use of animals as research subjects, as is sometimes advocated by some animal rights groups.

Below are provided some readings in the field of ethics. Your instructor may wish you to read one or more of the following. You might also look in various issues of the American Psychological Association's *Monitor*, a monthly magazine sent to every member of the American Psychological Association. In it you will often find articles addressing controversial issues in psychology. Excellent sources for recent articles are Medline and PsycLIT (Topic 2).

Ethics References:

Bowd, A. D. (1990). A decade of debate on animal research in psychology: Room for consensus? *Canadian Psychology, 31*, 74-82.

Christensen, L. (1988). Deception in psychological research: When is its use justified? *Personality and Social Psychology Bulletin, 14*, 664-675.

Coile, D.C., & Miller, N.E. (1984). How radical animal activists try to mislead humane people. *American Psychologist, 39*, 700-70l.

Hubbell, J.G. (1990). The "Animal Rights" war on medicine. *Reader's Digest*, June 1990, 70-76.

Jones, J.H. (1981). *Bad blood: The Tuskegee syphilis experiment*. New York: Free Press.

Leak, G.K. (1981). Student perception of coercion and value from participation in psychological research. *Teaching of Psychology, 8*, 147-149.

Makuch, R.W., & Johnson, M.F. (1989). Dilemmas in the use of active control groups in clinical research. *IRB: A Review of Human Subjects Research, 11*, 1-5.

Miller, N.E. (1983). Understanding the use of animals in behavioral research: Some critical issues. *Annals of the New York Academy of Sciences, 406*, 113-118.

National Academy of Sciences (1992). *Responsible science: Ensuring the integrity of the research process, Vol. 1* . Washington, D.C.: National Academy Press.

Novak, M.A., & Suomi, S.J. (1988). Psychological well-being of primates in captivity. *American Psychologist, 43*, 765-773.

Rowsell, H.C. (1988). The status of animal experimentation in Canada. *International Journal of Psychology, 23*, 377-381.

Sieber, J.E. (1982). Deception in social research I: Kinds of deception and the wrongs they may involve. *IRB: A Review of Human Subjects Research, 4*(9), 1-5.

Sieber, J.E. (1983). Deception in social research II: Evaluating the potential for wrong. *IRB: A Review of Human Subjects Research, 5*(1), 1-5.

Sieber, J. E. (199). *Planning ethically responsible research: A guide for students and internal review boards*. Newbury Park, CA: Sage Publications.

Sieber, J.E. (1983b). Deception in social research III: The nature and limits of debriefing . *IRB: A Review of Human Subjects Research, 5*(3), 1-4.

Smith, S.S., & Richardson, D. (1983). Amelioration of deception and harm in psychological research: The important role of debriefing. *Journal of Personality and Social Psychology, 44*, 1075-1082.

Ulrich, R. E. (1991). Animal rights, animal wrongs and the question of balance. *Psychological Science, 2*, 197-201.

LABORATORY EXERCISE 1: ETHICS OF PSYCHOLOGICAL RESEARCH

The purpose of this laboratory exercise is to help you become more sensitive to ethical issues through the examination of published psychological studies.

You should have read the chapter on ethics in your experimental methods book before proceeding with this assignment. In addition, you may wish to refer to some of the previous references as well as the following:

American Psychological Association (1982). *Ethical principles in the conduct of research with human participants*. Washington, D. C.: APA.

American Psychological Association (1990). Ethical principles of psychologists (Amended June 2, 1989). *American Psychologist, 45*, 390-395.

Dawkins, M. S., & Gosline, M. (Eds.) (1992). *Ethics in research on animal behaviour: Readings from "Animal Behaviour"* . San Diego, CA: Academic Press.

Guide for the care and use of laboratory animals. NIH Publication No. 85-23, Revised 1985, Office of Science and Health Reports, DRR/NIH, Betheseda, MD 20205.

The biomedical investigator's handbook for researchers using animal models. (1987). Washington, D.C.: Foundation for Biomedical Research.

The articles below were selected specifically to illustrate certain ethical issues that confront researchers. The abstract of the article is provided when possible. That these articles were selected does not mean that they have been judged to be unethical, although some researchers have raised questions about some of them since their publication. References to criticisms and rebuttals in the literature are also provided. Please remember that ethical issues are present in any study. Some of these studies were carried out before human subjects or animal care rights committees had been formed. In recent years our profession and society, in general, has become much more sensitive to ethical issues.

Your instructor will assign you one of two activities:

(1) Debate on ethics. You will be assigned to debate teams. Your debate team's responsibility will be to read the assigned set of articles. You are to prepare yourselves to argue either that there are or are not ethical problems involved, supporting your arguments with the ethical principles of the American Psychological Association and other sources. Once you have prepared your debate points, your instructor will have groups of students debate each of the assigned sets of research.

(2) Evaluation of ethical issues. You will be assigned articles under one or more of the sets of articles below. You are to write an evaluation of the ethical issues that are raised and the degree to which the rights and welfare of the subjects are protected. Indicate how you would resolve any ethical issues.

ANIMAL RESEARCH:

A. Autonomic nervous system learning

1. DiCara, L. V., & Miller, N. E. (1968). Changes in heart rate instrumental-learned by curarized rats as avoidance responses. *Journal of Comparative and Physiological Psychology, 65*, 8-12.

2 groups of curarized rats learned to increase or decrease, respectively, their heart rates in order to escape or avoid mild electric shocks. Responses in the appropriate direction were greater during

the stimulus preceding shock than during control intervals between shock; they change in the opposite direction, toward the initial pretraining level, during the different stimulus preceding nonshock. Electromyograms indicated complete paralysis of the gastrocnemius muscle throughout training and for a period of at least 1 hr. thereafter (p. 8).

2. Di Cari, L. V. (1970). Learning in the autonomic nervous system. *Scientific American, 222,* 30-39.

B. Learned Helplessness

1. Seligman, M. E. P., & Beagley, G. (1975). Learned helplessness in the rat. *Journal of Comparative and Physiological Psychology, 88,* 534-541.

Four experiments attempted to produce behavior in the rat parallel to the behavior characteristic of learned helplessness in the dog. When rats received escapable, inescapable, or no shock and were later tested in jump-up escape, both inescapable and no-shock controls failed to escape. When bar pressing, rather than jumping up, was used as the tested escape response, fixed ratio (FR) 3 was interfered with by inescapable shock, but not lesser ratios. With FR-3, the no-shock control escaped well. Interference with escape was shown to be a function of the inescapability of shock and not shock per se: Rats that were "put through" and learned a prior jump-up escape did not become passive, but their yoked, inescapable partners did. Rats, as well as dogs, fail to escape shock as a function of prior inescapability, exhibiting learned helplessness (p.534).

2. Seligman, M. E. P., & Groves, D. P. (1970). Nontransient learned helplessness. *Psychonomic Science, 19,* 191-192.

Dogs who receive repeated, spaced exposure to inescapable electric shock in a Pavlovian hammock fail to escape shock in a shuttlebox 1 week later, while one session of inescapable shock produces only transient interference. Cage-raised beagles are more susceptible to interference produced by inescapable shock than are mongrels of unknown history. These results are compatible with learned helplessness and contradict the hypothesis that failure to escape shock is produced by transient stress (p. 191).

HUMAN RESEARCH:

A. Portrayals of sexual violence

1. Malamuth, N. M., Heim, M., & Feshbach, S. (1980). Sexual responsiveness of college students to rape depictions: Inhibitory and disinhibitory effects. *Journal of Personality and Social Psychology, 38,* 399-408.

Two experiments were conducted to identify the specific dimensions in portrayals of sexual violence that inhibit or disinhibit the sexual responsiveness of male and female college students. The first experiment replicated earlier findings that normals are less sexually aroused by portrayals of sexual assault than by depictions of mutually consenting sex. In the second experiment, it was shown that portraying the rape victim as experiencing an involuntary orgasm disinhibited subjects' sexual responsiveness and resulted in levels of arousal comparable to those elicited by depictions of mutually consenting sex. Surprisingly, however, it was found that although female subjects were most aroused when the rape victim was portrayed as experiencing an orgasm and no pain, males were most aroused when the victim experienced an orgasm and pain. The relevance of these data to pornography and to the common belief among rapists that their victims derive pleasure from being assaulted is discussed. Misattribution, identification, and power explanations of the findings are also discussed. Finally, it is suggested that arousing stimuli that fuse sexuality and violence may have antisocial effects (p. 399).

2. Sherif, C. W. (1980).Comment on ethical issues in Malamuth, Heim, and Feshbach's "Sexual responsiveness of college students to rape depictions: Inhibitory and disinhibitory effects". *Journal of Personality and Social Psychology, 38*, 409-412.

3. Malamuth, N. M., Feshbach, S., & Heim, M. (1980). Ethical issues and exposure to rape stimuli: A reply to Sherif. *Journal of Personality and Social Psychology, 38*, 413-415.

4. Check, J. V., & Malamuth, N. M. (1984). Can there be positive effects of participation in pornography experiments. *Journal of Sex Research, 20*, 14-31.

Conducted a 2-phase experiment in response to recent ethical concerns about the possible antisocial effects of exposing research Ss to pornographic rape portrayals. In Phase 1, 64 male and 64 female undergraduates were randomly assigned to read either an "acquaintance" or a "stranger" rape depiction, or to read control materials. Ss who read the rape depictions were then given a rape debriefing that included a communication about the undesirable desensitizing effects of pairing sexual violence with other highly explicit and pleasing sexual stimuli. Half of the Ss who read the control materials were also given the rape debriefing. In Phase 2, Ss were presented with a number of newspaper articles in which a newspaper report of a rape was embedded and asked to give their opinions. Results indicate that the rape debriefing generally increased Ss' perceptions of pornography as a cause of rape. Ss in the rape debriefing conditions also gave the rapist in the newspaper report a higher sentence and saw the rape victim as less responsible than did Ss in the control conditions. This latter effect, however, occurred only under conditions where Ss had earlier been exposed to an example of a rape depiction that was relevant to both the rape myths discussed in the rape debriefing and the newspaper report of the rape (p. 14).

B. Participation in a burglary

1. West, S. G., Gunn, S. P., & Chernicky, P. (1975). Ubiquitous Watergate: An attributional analysis. *Journal of Personality and Social Psychology, 32*, 55-65.

Actor-observer differences in causal attribution were investigated in an experiment involving two separate studies. Study 1 was a field experiment in which 80 subjects (actors) were presented with elaborate plans for burglarizing a local advertising firm under one of four experimental conditions: (a) a control condition, (b) $2,000 (reward), (c) government sponsorship but no immunity from prosecution, and (d) government sponsorship plus immunity. In Study 2, 238 subjects (observers) read a description of a student agreeing or refusing to participate in the burglary under one of the four experimental conditions. Consistent with Jones and Nisbett's 1971 theory, actors made more environmental attributions, while observers made more dispositional attributions. Further, observers made more dispositional attributions when the actor agreed than when he refused, except in the reward condition, where this relationship was reversed. The results are interpreted with reference to the disparate explanations of Watergate offered by the Nixon administration and the press (p. 55).

2. Cook, Stuart W. (1975). A comment on the ethical issues involved in West, Gunn, and Chernicky's "Ubiquitous Watergate: An attributional analysis". *Journal of Personality and Social Psychology, 32*, 66-68.

C. Behavioral study of obedience in children

1. Shanab, M. E., & Yahya, K. (1977). A behavioral study of obedience in children. *Journal of Personality and Social Psychology, 35*, 530-536.

Using Milgram's original test of obedience, 192 Jordanian subjects were tested in a 2 X 2 X 3 design in which sex, two kinds of punishment instructions, and three levels of age groups (6-8, 10-12, 14-16 years) were combined factorially. The instructions issued to the experimental group were identical to those used in Milgram's paradigm in that teacher subjects were asked to administer shock to confederate learners each time the latter made a mistake in a paired-associate task and to increase the shock level with each additional mistake. The subjects in the control group were given a free choice of delivering or not delivering shock each time the learner made a mistake. The results showed that 73% of all experimental subjects, as opposed to 16% of the control subjects, continued to deliver shock to the end of the shock scale. Neither age nor sex differences in obedience rate were found. However, significantly more obedient females than males reported that they punished the learners because they were obeying orders (p. 530).

D. Invasion of personal space

1. Middlemist, R. D., Knowles, E. S., & Matter, C. (1976). Personal space invasions in the lavatory: Suggestive evidence for arousal. *Journal of Personality and Social Psychology, 33*, 541-546.

The hypothesis that personal space invasions produce arousal was investigated in a field experiment. A men's lavatory provided a setting where norms for privacy were salient, where personal space invasions could occur in the case of men urinating, where the opportunity for compensatory responses to invasion were minimal, and where proximity-induced arousal could be measured. Research on micturition indicates that social stressors inhibit relaxation of the external urethral sphincter, which would delay the onset of micturition, and that they increase intravesical pressure, which would shorten the duration of micturition once begun. Sixty lavatory users were randomly assigned to one of three levels of interpersonal distance and their micturition times were recorded. In a three-urinal lavatory, a confederate stood immediately adjacent to a subject, one urinal removed, or was absent. Paralleling the results of a correlational pilot study, close interpersonal distances increased the delay of onset and decreased the persistence of micturition. These findings provide objective evidence that personal space invasions produce physiological changes associated with arousal.

E. Treating autism and severe behavior disorders

1. Lovaas, O. I., Schaeffer, B., & Simmons, J. Q. (1965). Experimental studies in childhood schizophrenia: Building social behavior in autistic children by use of electric shock. *Journal of Experimental Research in Personality, 1*, 99-109.

2. Lovaas, O. I., & Favell, J. E. (1987). Protection for clients undergoing aversive/restrictive interventions. Special Issue: New developments in the treatment of persons exhibiting autism and severe behavior disorders. *Education and Treatment of Children, 10*, 311-325.

3. Lovaas, O. I. (1987). Behavioral treatment and normal educational and intellectual functioning in young autistic children. *Journal of Consulting and Clinical Psychology, 55*, 3-9.

4. Schopler, E. (1988). Concerns about misinterpretation and uncritical acceptance of exaggerated claims. *American Psychologist, 43*, 658.

5. Lovaas, O. I., Smith, T., & McEachin, J. J. (1989). Clarifying comments on young autism study: Reply to Schopler, Short, and Mesibov. *Journal of Consulting and Clinical Psychology, 57*, 165-167.

6. Lovaas, O. I. (1989). Concerns about misinterpretation and placement of blame. *American Psychologist, 44*, 143-144.

7. Feguson, L. R. (1978). The competence and freedom of children to make choices regarding participation in research: A statement. *Journal of Social Issues, 34*, 114-121.

LABORATORY EXERCISE 2: EVALUATING AND DEVELOPING CONSENT FORMS

The purpose of this laboratory exercise is to give you experience in developing informed consent forms.

First read the following examples of informed consent and then develop your own consent form for one of the experiments or surveys you carried out in your class this year, or an experiment you would like to carry out. Obtain the necessary forms and directions from your department's Human Subjects Committee or from the Institutional Review Board. Prepare the consent form as if you were going to present it to an Institutional Review Board for approval. Turn this into your instructor.

Listed below are some research articles that investigated the impact of consent forms on research as well as some articles that address issues surrounding informed consent regulations. You may be asked to read one or more, and present a summary to your classmates.

References:

Bradley, E.J., & Lindsay, R.C. (1987). Methodological and ethical issues in child abuse research. *Journal of Family Violence, 2*, 239-255.

Cassel, C.K. (1987). Informed consent for research in geriatrics: History and concepts. *Journal of the American Geriatrics Society, 35*, 542-544.

Christensen, L. (1988). Deception in psychological research: When is its use justified? *Personality and Social Psychology Bulletin, 14*, 664-675.

Finney, P.D. (1987). When consent information refers to risk and deception: Implications for social research. *Journal of Social Behavior and Personality, 2*, 37-48.

Korn, J.H. (1988). Students' roles, rights, and responsibilities as research participants. *Teaching of Psychology, 15*, 74-78.

Lavelle, J. C., Byrne, D. J., Rice, P., & Cuschieri, A. (1993). Factors affecting quality of informed consent. *British Medical Journal, 306*, 885-890.

Lindsay, R.C., & Holden, R.R. (1987). The introductory psychology subject pool in Canadian universities. *Canadian Psychology, 28*, 45-52.

Mahler, D.M. (1986). When to obtain informed consent in behavioral research: A study of mother-infant bonding. *IRB: A Review of Human Subjects Research, 8*, 7-11.

Newton, L.H. (1984). Agreement to participate in research: Is that a promise? *IRB: A Review of Human Subjects Research, 6*, 7-9.

Pattullo, E.L. (1987). Exemption from review, not informed consent. *IRB: A Review of Human Subjects Research, 9*, 6-8.

Sieber, J.E., & Saks, M.J. (1989). A census of subject pool characteristics and policies. *American Psychologist, 44*, 1053-1061.

Sieber, J.E., & Stanley, B. (1988). Ethical and professional dimensions of socially sensitive research. *American Psychologist, 43*, 49-55.

Trice, A.D. (1986). Informed consent: III. Informing subjects that performance will not affect credit may affect performance. *Perceptual and Motor Skills, 62*, 178.

Trice, A.D. (1987). Informed consent: IV. The effects of the timing of giving consent on experimental performance. *Journal of General Psychology, 114*, 125-128.

Trice, A.D., & Ogden, E.P. (1986). Informed consent: I. The institutional non-liability clause as a liability in recruiting research subjects. *Journal of Social Behavior and Personality, 1*, 391-396.

Trice, A.D., & Ogden, E.P. (1987). Informed consent: IX. Effects of the withdrawal clause in longitudinal research. *Perceptual and Motor Skills, 65*, 135-138.

Young, D. R., Hooker, D. T., Freeberg, F. E. (1990). Informed consent documents: Increasing comprehension by reducing reading level. *IRB: A Review of Human Subjects Research, 12*, 1-5.

INSTRUMENT TO OBTAIN INFORMED CONSENT (Experimental Group)

You are invited to participate in a study investigating the effect of a dietary change in treating the premenstrual syndrome. The premenstrual syndrome is a term that refers to a group of mood, behavioral and/or physical symptoms such as tension, irritability, mood swings, bloating and carbohydrate cravings that occur regularly with the menstrual cycle. One of the many treatments that has been advocated is a dietary change. Although a dietary change is frequently advocated, it has never been tested to determine if it produces any noticeable benefit. This study is being conducted to determine if a dietary change can eliminate or reduce the severity of the premenstrual syndrome.

PROCEDURE TO BE FOLLOWED IN THE STUDY: To accomplish the goals of the study, we will ask you to complete a daily calendar of the symptoms you experience during a complete menstrual cycle. You will then be asked to undergo a physical examination and complete several psychological tests as well as a structured interview to determine if you meet the stringent criteria we are using for defining the premenstrual syndrome or PMS. If you meet the criteria we have established, you will be asked to come to the Human Psychophysiological Research Laboratory one evening during the beginning and one evening toward the end of your menstrual cycle. Each of these evenings you will provide a blood sample and be asked to complete several psychological tests prior to and each half-hour for two hours after eating a high carbohydrate meal. Following the completion of this assessment, you will be randomly assigned to one or two dietary treatment groups and given a set of dietary instructions to follow for the next month. During this month we will again ask you to maintain a daily calendar of all symptoms experienced. During this time we will also ask you to maintain a record of everything you eat and drink and to provide a saliva sample so that we can be sure that you are following the diet.

DISCOMFORTS AND RISKS FROM PARTICIPATING IN THE STUDY: We will try to make the study as comfortable as possible. However, you may experience some symptoms such as headaches during the first few days while following the diet. These will disappear within several days. There are no risks to participating in the study that we have identified other than the remote possibility of infection or bruising resulting from drawing the blood sample.

ALTERNATIVE PROCEDURES APPROPRIATE TO YOU: There are several alternative procedures that have been used for treating PMS such as progesterone, diuretics, antidepressants, and relaxation training. However, good scientific evidence documenting the benefits of these therapies does not exist.

EXPECTED BENEFITS: We expect that one of the two dietary alterations we are investigating will result in an elimination or a reduction in the experience of the PMS. If you are a student in Introductory Psychology, you can receive credit for participating in a psychological experiment by participating in this study.

CONFIDENTIALITY OF THE RESULTS: The results of the study will be kept strictly confidential. At no time will we release the results of the study to anyone other than individuals working on the project without your written consent.

FINANCIAL COMPENSATION AND MEDICAL TREATMENT: Although we do not anticipate any possibility of physical injury, you should know that there is no free medical treatment or compensation of any kind available.

FREEDOM TO WITHDRAW: You are free to withdraw your consent to participate and discontinue participation in the project at any time.

WAIVE OR RELEASE OF THE INSTITUTION: You affirm that you have not been requested to waive, or to release the institution or its agents from liability for negligence and that you have read and understand the above information and have been given a chance to ask questions as well as have questions answered.

_____   _____
Date   **Signature of Participant**

<u>**Contact Person for Additional Information**</u>: Larry Christensen, Ph.D., 409-845-2549

_____, **Chair of IRB**

INSTRUMENT TO OBTAIN INFORMED CONSENT (control subjects)

You are invited to participate in a study investigating the effect of a dietary change in treating the premenstrual syndrome. The premenstrual syndrome is a term that refers to a group of mood, behavioral and/or physical symptoms such as tension, irritability, mood swings, bloating and carbohydrate cravings that occur regularly with the menstrual cycle. One of the many treatments that has been advocated is a dietary change. Although a dietary change is frequently advocated, it has never been tested to determine if it produces any noticeable benefit. This study is being conducted to determine if a dietary change can eliminate or reduce the severity of the premenstrual syndrome. Your participation is needed to assess the effect of the dietary alteration on individuals with an absence of PMS symptoms.

PROCEDURE TO BE FOLLOWED IN THE STUDY: To accomplish the goals of the study, we will ask you to complete a daily calendar of the symptoms you experience during a complete menstrual cycle. You will then be asked to undergo a physical examination and complete several psychological tests as well as a structured interview to determine if you meet the stringent criteria we are using for defining absence of the premenstrual syndrome or PMS. If you meet the criteria we have established, you will be asked to come to the Human Psychophysiological Research Laboratory one evening during the beginning and one evening toward the end of your menstrual cycle. Each of these evenings you will be asked to complete several psychological tests prior to and each half-hour for two hours after eating a high carbohydrate meal and have a blood sample drawn.

DISCOMFORTS AND RISKS FROM PARTICIPATING IN THE STUDY: We will try to make the study as comfortable as possible. The only risk that exists from participating in the study is the remote possibility of infection and the possibility of bruising resulting from drawing the blood sample.

ALTERNATIVE PROCEDURES APPROPRIATE TO YOU: The alternative procedure available to you is to decline participation.

EXPECTED BENEFITS: There are no anticipated benefits which you can expect to personally derive from participation. However, your participating will assist us in identifying the benefit of a dietary alteration in treatment of PMS. If you are a student in Introductory Psychology, you can receive credit for participating in a psychological experiment by participating in this study.

CONFIDENTIALITY OF THE RESULTS: The results of the study will be kept strictly confidential. At no time will we release the results of the study to anyone other than individuals working on the project without your written consent.

FINANCIAL COMPENSATION AND MEDICAL TREATMENT: Although we do not anticipate any possibility of physical injury, you should know that there is no free medical treatment or compensation of any kind available.

FREEDOM TO WITHDRAW: You are free to withdraw your consent to participate and discontinue participation in the project at any time.

WAIVE OR RELEASE OF THE INSTITUTION: You affirm that you have not been requested to waive, or to release the institution or its agents from liability for negligence and that you have read and understand the above information and have been given a chance to ask questions as well as have questions answered.

_____   _____
Date   Signature of Participant

<u>Contact Person for Additional Information</u>: Larry Christensen, Ph.D., 409-845-2549

_____, Chair of IRB

CONSENT FORM: ANONYMOUS SURVEY

This is a COMPLETELY ANONYMOUS survey about the attitudes and activities of the general UNDERGRADUATE STUDENT AT VIRGINIA TECH on the topics of alcohol usage, sexual practices, condom use, and attitudes toward AIDS. It is being conducted by undergraduates in a psychology course that is devoted to learning about research methods. This questionnaire has been approved for administration by the Human Subjects Committee of the Psychology Department and by the Institutional Review Board at Virginia Tech.

Since some of the material is personal, let us emphasize that this is completely anonymous and you cannot be identified in any way. You are not to put your name on the form. To further ensure anonymity you are to fill out the questionnaire in such a manner that the experimenter cannot see your responses. Finally, you are to place your questionnaire through a slit in a sealed manila envelope. The student researcher has signed a code of ethics indicating that he/she will not examine the individual response forms that are gathered. The response forms will be gathered all together from the class and submitted to the testing center on campus for transfer to a summary data set.

If you agree to participate, you will be asked to complete questions on an opscan. You will be asked to answer questions concerning your recent drinking and sexual habits as well as your opinions about AIDS. YOU MAY REFRAIN FROM ANSWERING ANY QUESTION.

As part of the class activity, this information will be coded into the computer and then analyzed. We hope to present the results to the Dean of Students at Virginia Tech and to have a summary published in the student newspaper. In addition, a summary of the findings will be available from Dr. Crawford, Psychology Department, in April, 1993.

Should you have any questions, please feel free to contact any of the following individuals:

Helen Crawford, Ph.D., Psychology Faculty Supervisor	231-6520
Joseph Franchina, Ph.D., Chair, Human Subjects Committee	231-6581
Janet Johnson, Ph.D., Chair, Institutional Review Board	231-9359

Please proceed to answer the questionnaire, following its directions, if you agree to participate. Remember to remove yourself from the experimenter and your friends so that you answer the questionnaire out of their vision. Then place the questionnaire into the envelope.

CODE OF ETHICS

I, _____ , agree to follow the ethical guidelines of the American Psychological Association. I will not examine any of the collected questionnaires.

Signed: _____   Date: _____

LABORATORY EXERCISE 3: DEVELOPING A DEBRIEFING FORM

The purpose of this laboratory exercise is to give you experience in developing a debriefing form. At the conclusion of every experiment it is incumbent upon the experimenter to debrief the subject. This debriefing should inform the subject of the purpose of the experiment, the procedure followed in accomplishing the purpose of the experiment, the experimental condition the subject participated in and eliminate any unacceptable aspects created by the experiment. One aspect which has been considered unacceptable by some is deception. If an experiment contains deception the debriefing should inform the subject of the nature of the deception, why it was necessary, and how it was accomplished. Emphasis should be placed on the fact that deception was necessary to accomplish the purpose of the experiment and not on the fact that the subject was duped (Christensen, 1988).

On the next page is an example of a debriefing form which was used in an actual experiment on social interaction (Simpson & Gangested, 1990). Read this example and then develop your own debriefing form for one of the experiments you carried out in your class this year, or an experiment you would like to carry out.

References:

Christensen, L. (1988). Deception in psychological research: When is its use justified? *Personality and Social Psychology Bulletin, 14*, 664-675.

Kelman, H. C. (1967). Human use of subjects: The problem of deception in social psychological experiments. *Psychological Bulletin, 67*, 1-11.

Sieber, J. E. (1983). Deception in social research III: The nature and limits of debriefing. *IRB: A Review of Human Subjects Research, 5*, 1-4.

Simpson, J. A., & Gangested, S. W. (1990). *Relationship initiation: The dating game revisited*. Paper presented at the International Conference on Close Relationships, Nags Head, North Carolina.

Smith, S. S., & Richardson, D. (1983). Amelioration of deception and harm in psychological research: The important role of debriefing. *Journal of Personality and Social Psychology, 44*, 1075-1082.

DEBRIEFING FORM

You have just participated in research on social interaction conducted by researchers within the Department of Psychology at Texas A&M University.

This research is designed to determine whether different "styles" of attachment to others influence the nature of spontaneous interaction that transpires between dating couples.

Past research has identified 3 primary attachment "styles". One style is characterized by a tendency to typically seek out a moderate amount of comfort from a romantic partner when one is experiencing stress or anxiety. A second style involves a tendency to seek out a minimal amount of comfort when one is under stress or anxiety. Finally, the third style entails a tendency to seek out a great amount of comfort when one is experiencing stress or anxiety. All 3 styles reflect _normal_ adaptations to one's social environment.

In this study, we first had you indicate (on the self-report questionnaire survey) which style best characterizes you. We then videotaped the interaction that transpired between you and your partner as you waited for the "experimental" portion of the study to begin. Following this, we told one of you that you would be experiencing a stressful/anxiety-provoking event as part of the experimental portion of the study. We then videotaped your interaction during the 5 minute waiting period prior to this stressful event (which, of course, you never experienced). Finally, after the experimental informed you that a "scheduling problem" prohibited the study from continuing (which was a cover story used to conclude the study), we videotaped your interaction to see how the two of you behaved once the "stressor" was removed. We hypothesize that individuals who possess different attachment styles will have very different types of spontaneous interactions with their partners across the 3 interaction episodes.

To successfully study the spontaneous, natural, and uninhibited interaction that occurs in romantic relationships, we had to covertly videotape your spontaneous interactions. We did not warn you about this videotaping because previous research has shown that when you inform people that their behavior is being monitored, they alter it in very constrained and unnatural ways. In particular, their natural interaction tendencies are inhibited and distorted.

Since we are interested in _naturalistic_ interaction, we could not inform you about this procedure beforehand. However, _if for any reason you do not want your videotaped interaction to be evaluated by coders, we will be more than happy to erase the videotape._ Moreover, you will _not_ be penalized in any way for doing so.

To ensure confidentiality, each participant will be given anonymous identification numbers. Your data will be coded by 3 independent researchers who will not know you. _After the coding is completed (within about 6 months), the videotapes will be erased_. This procedure will guarantee that your name will never be associated with your interaction episodes.

If you have any additional questions about this research, please contact either Dr. Jeff Simpson or Dr. Steve Rholes at 845-7146.

I hereby grant permission to allow the videotape to be used only for research purposes, after which it will be destroyed.

_____
Signature

NOTE: This example was supplied by Jeff Simpson, Department of Psychology, Texas A&M University.

APPENDIX A

Surveys for Topic 6:

Laboratory Exercise 2: Survey on Alcohol Use (pp. 57-60)

Laboratory Exercise 3: Survey on Eating Disorders (pp. 61-62)

Laboratory Exercise 4: Survey on ESP Attitudes and Experiences (pp. 63-64)

ANONYMOUS SURVEY: ALCOHOL USAGE BY COLLEGE STUDENTS

Gender: ____ male ____ female   Age now: _____   Age at first drink: _____

Year in college: __ freshman   __ sophomore   __ junior   __ senior   __ grad student

Since school started, on the average, how often do you drink alcohol?
___ every day
___ at least once a week, not every day
___ at least once a month, not every week
___ more than once a year, less than once a month
___ once a year
___ never

Since school started, on the average, how much alcohol do you drink at any one time?
___ more than 6 alcoholic drinks
___ 5 - 6 alcoholic drinks
___ 3 - 4 alcoholic drinks
___ 1 - 2 alcoholic drinks
___ less than 1 alcoholic drink
___ I do not drink

Check appropriate responses:

After drinking alcohol, I	Once or more this past year	Prior to this year	Never
had a hangover	_____	_____	_____
felt nauseated or vomited	_____	_____	_____
experienced a blackout where I could not remember incidents or conversations the next day	_____	_____	_____
binged on alcohol for two or more days	_____	_____	_____
drove a car after several drinks of alcohol	_____	_____	_____
was arrested for DUI	_____	_____	_____
went to class after several drinks of alcohol	_____	_____	_____
missed class or work because of drinking or hangover	_____	_____	_____
was criticized for drinking too much	_____	_____	_____
thought I might have a drinking problem	_____	_____	_____
got a lower grade than I would have if I had not drunk	_____	_____	_____
got into a fight	_____	_____	_____
damaged property	_____	_____	_____
got in trouble with the law	_____	_____	_____

ANONYMOUS SURVEY: ALCOHOL USAGE BY COLLEGE STUDENTS

Gender: ____ male ____ female   Age now: ____   Age at first drink: ____

Year in college: __ freshman   __ sophomore   __ junior   __ senior   __ grad student

Since school started, on the average, how often do you drink alcohol?	Since school started, on the average, how much alcohol do you drink at any one time?
___ every day	___ more than 6 alcoholic drinks
___ at least once a week, not every day	___ 5 - 6 alcoholic drinks
___ at least once a month, not every week	___ 3 - 4 alcoholic drinks
___ more than once a year, less than once a month	___ 1 - 2 alcoholic drinks
___ once a year	___ less than 1 alcoholic drink
___ never	___ I do not drink

Check appropriate responses:

After drinking alcohol, I	Once or more this past year	Prior to this year	Never
had a hangover	_____	_____	_____
felt nauseated or vomited	_____	_____	_____
experienced a blackout where I could not remember incidents or conversations the next day	_____	_____	_____
binged on alcohol for two or more days	_____	_____	_____
drove a car after several drinks of alcohol	_____	_____	_____
was arrested for DUI	_____	_____	_____
went to class after several drinks of alcohol	_____	_____	_____
missed class or work because of drinking or hangover	_____	_____	_____
was criticized for drinking too much	_____	_____	_____
thought I might have a drinking problem	_____	_____	_____
got a lower grade than I would have if I had not drunk	_____	_____	_____
got into a fight	_____	_____	_____
damaged property	_____	_____	_____
got in trouble with the law	_____	_____	_____

ANONYMOUS SURVEY: ALCOHOL USAGE BY COLLEGE STUDENTS

Gender: ____ male ____ female   Age now: _____   Age at first drink: _____

Year in college: ____ freshman ____ sophomore ____ junior ____ senior ____ grad student

Since school started, on the average, how often do you drink alcohol?
- ____ every day
- ____ at least once a week, not every day
- ____ at least once a month, not every week
- ____ more than once a year, less than once a month
- ____ once a year
- ____ never

Since school started, on the average, how much alcohol do you drink at any one time?
- ____ more than 6 alcoholic drinks
- ____ 5 - 6 alcoholic drinks
- ____ 3 - 4 alcoholic drinks
- ____ 1 - 2 alcoholic drinks
- ____ less than 1 alcoholic drink
- ____ I do not drink

Check appropriate responses:

After drinking alcohol, I	Once or more this past year	Prior to this year	Never
had a hangover	____	____	____
felt nauseated or vomited	____	____	____
experienced a blackout where I could not remember incidents or conversations the next day	____	____	____
binged on alcohol for two or more days	____	____	____
drove a car after several drinks of alcohol	____	____	____
was arrested for DUI	____	____	____
went to class after several drinks of alcohol	____	____	____
missed class or work because of drinking or hangover	____	____	____
was criticized for drinking too much	____	____	____
thought I might have a drinking problem	____	____	____
got a lower grade than I would have if I had not drunk	____	____	____
got into a fight	____	____	____
damaged property	____	____	____
got in trouble with the law	____	____	____

ANONYMOUS SURVEY: ALCOHOL USAGE BY COLLEGE STUDENTS

Gender: ____ male ____ female   Age now: _____   Age at first drink: _____

Year in college: __ freshman   __ sophomore   __ junior   __ senior   __ grad student

Since school started, on the average, how often do you drink alcohol?	Since school started, on the average, how much alcohol do you drink at any one time?
___ every day	___ more than 6 alcoholic drinks
___ at least once a week, not every day	___ 5 - 6 alcoholic drinks
___ at least once a month, not every week	___ 3 - 4 alcoholic drinks
___ more than once a year, less than once a month	___ 1 - 2 alcoholic drinks
___ once a year	___ less than 1 alcoholic drink
___ never	___ I do not drink

Check appropriate responses:

After drinking alcohol, I	Once or more this past year	Prior to this year	Never
had a hangover	_____	_____	_____
felt nauseated or vomited	_____	_____	_____
experienced a blackout where I could not remember incidents or conversations the next day	_____	_____	_____
binged on alcohol for two or more days	_____	_____	_____
drove a car after several drinks of alcohol	_____	_____	_____
was arrested for DUI	_____	_____	_____
went to class after several drinks of alcohol	_____	_____	_____
missed class or work because of drinking or hangover	_____	_____	_____
was criticized for drinking too much	_____	_____	_____
thought I might have a drinking problem	_____	_____	_____
got a lower grade than I would have if I had not drunk	_____	_____	_____
got into a fight	_____	_____	_____
damaged property	_____	_____	_____
got in trouble with the law	_____	_____	_____

ANONYMOUS SURVEY: ALCOHOL USAGE BY COLLEGE STUDENTS

Gender: ____ male ____ female   Age now: _____   Age at first drink: _____

Year in college: __ freshman   __ sophomore   __ junior   __ senior   __ grad student

Since school started, on the average, how often do you drink alcohol?
___ every day
___ at least once a week, not every day
___ at least once a month, not every week
___ more than once a year, less than once a month
___ once a year
___ never

Since school started, on the average, how much alcohol do you drink at any one time?
___ more than 6 alcoholic drinks
___ 5 - 6 alcoholic drinks
___ 3 - 4 alcoholic drinks
___ 1 - 2 alcoholic drinks
___ less than 1 alcoholic drink
___ I do not drink

Check appropriate responses:

After drinking alcohol, I	Once or more this past year	Prior to this year	Never
had a hangover	_____	_____	_____
felt nauseated or vomited	_____	_____	_____
experienced a blackout where I could not remember incidents or conversations the next day	_____	_____	_____
binged on alcohol for two or more days	_____	_____	_____
drove a car after several drinks of alcohol	_____	_____	_____
was arrested for DUI	_____	_____	_____
went to class after several drinks of alcohol	_____	_____	_____
missed class or work because of drinking or hangover	_____	_____	_____
was criticized for drinking too much	_____	_____	_____
thought I might have a drinking problem	_____	_____	_____
got a lower grade than I would have if I had not drunk	_____	_____	_____
got into a fight	_____	_____	_____
damaged property	_____	_____	_____
got in trouble with the law	_____	_____	_____

THE BULIT-R[1]

Answer each question by circling the appropriate answer. Please respond to each item as honestly as possible; remember all of the information you provide is anonymous.

1. I am satisfied with my eating patterns.
   a. agree
   b. neutral
   c. disagree a little
   d. disagree
   e. disagree strongly

2. Would you presently call yourself a "binge eater"?
   a. yes, absolutely
   b. yes
   c. yes, probably
   d. yes, possibly
   e. no, probably not

3. Do you feel you have control over the amount of food you consume?
   a. most or all of the time
   b. a lot of the time
   c. occasionally
   d. rarely
   e. never

4. I am satisfied with the shape and size of my body.
   a. frequently or always
   b. sometimes
   c. occasionally
   d. rarely
   e. seldom or never

5. When I feel that my eating behavior is out of control, I try to take rather extreme measures to get back on course (strict dieting, fasting, laxatives, diuretics, self-induced vomiting, or vigorous exercise).
   a. always
   b. almost always
   c. frequently
   d. sometimes
   e. never or my eating behavior is never out of control

6. I am obsessed about the size and shape of my body.
   a. always
   b. almost always
   c. frequently
   d. sometimes
   e. seldom or never

7. There are times when I rapidly eat a very large amount of food.
   a. more than twice a week
   b. twice a week
   c. once a week
   d. 2-3 times a month
   e. once a month or less (or never)

---

[1]Revised with permission of M. Thelen.

8. How long have you been binge eating (eating uncontrollably to the point of stuffing yourself)?
    a. not applicable; I don't binge eat
    b. less than 3 months
    c. 3 months - 1 year
    d. 1 - 3 years
    e. 3 or more years

9. Most people I know would be amazed if they knew how much food I can consume at one sitting.
    a. without a doubt
    b. very probably
    c. probably
    d. possibly
    e. no

10. Compared with others your age, how preoccupied are you about your weight and body shape?
    a. a great deal more than average
    b. much more than average
    c. more than average
    d. a little more than average
    e. average or less than average

11. I am afraid to eat anything for fear that I won't be able to stop.
    a. always
    b. almost always
    c. frequently
    d. sometimes
    e. seldom or never

12. I feel tormented by the idea that I am fat or might gain weight.
    a. always
    b. almost always
    c. frequently
    d. sometimes
    e. seldom or never

13. How often do you intentionally vomit after eating?
    a. 2 or more times a week
    b. once a week
    c. 2-3 times a month
    d. once a month
    e. less than once a month or never

14. I eat a lot of food when I'm not even hungry.
    a. very frequently
    b. frequently
    c. occasionally
    d. sometimes
    e. seldom or never

15. My eating patterns are different from the eating patterns of most people.
    a. always
    b. almost always
    c. frequently
    d. sometimes
    e. seldom or never

16. After I binge eat I turn to one of several strict methods to try to keep from gaining weight (vigorous exercise, strict dieting, fasting, self-induced vomiting, laxatives, or diuretics).
    a. never or I don't binge eat
    b. rarely
    c. occasionally
    d. a lot of the time
    e. most or all of the time

17. When engaged in an eating binge, I tend to eat foods that are high in carbohydrates (sweets and starches).
    a. always
    b. almost always
    c. frequently
    d. sometimes
    e. seldom, or I don't binge

18. Compared to most people, my ability to control my eating behavior seems to be:
    a. greater than others' ability
    b. about the same
    c. less
    d. much less
    e. I have absolutely no control

19. I would presently label myself a 'compulsive eater' (one who engages in episodes of uncontrolled eating).
    a. absolutely
    b. yes
    c. yes, probably
    d. yes, possibly
    e. no, probably not

20. I hate the way my body looks after I eat too much.
    a. seldom or never
    b. sometimes
    c. frequently
    d. almost always
    e. always

21. When I am trying to keep from gaining weight, I feel that I have to resort to vigorous exercise, strict dieting, fasting, self-induced vomiting, laxatives, or diuretics.
    a. never
    b. rarely
    c. occasionally
    d. a lot of the time
    e. most of all of the time

22. Do you believe that it is easier for you to vomit than it is for most people?
    a. yes, it's no problem at all for me
    b. yes, it's easier
    c. yes, it's a little easier
    d. about the same
    e. no, it's less easy

23. I feel that food controls my life.
    a. always
    b. almost always
    c. frequently
    d. sometimes
    e. seldom or never

24. When consuming a large quantity of food, at what rate of speed do you usually eat?
    a. more rapidly than most people have ever eaten in their lives
    b. a lot more rapidly than most people
    c. a little more rapidly than most people
    d. about the same rate as most people
    e. more slowly than most people (or not applicable)

25. Right after I binge eat I feel:
    a. so fat and bloated I can't stand it
    b. extremely fat
    c. fat
    d. a little fat
    e. OK about how my body looks or I never binge eat

26. Compared to other people of my sex, my ability to always feel in control of how much I eat is:
    a. about the same or greater
    b. a little less
    c. less
    d. much less
    e. a great deal less

27. In the last 3 months, on the average how often did you binge eat (eat uncontrollably to the point of stuffing yourself)?
    a. once a month or less (or never)
    b. 2-3 times a month
    c. once a week
    d. twice a week
    e. more than twice a week

28. Most people I know would be surprised at how fat I look after I eat a lot of food.
    a. yes, definitely
    b. yes
    c. yes, probably
    d. yes, possibly
    e. no, probably not or I never eat a lot of food

29. What gender are you?
    a. male
    b. female

30. What age are you?
    a. Under 18
    b. 18 - 20
    c. 21 - 25
    d. 26 or older

THE BULIT-R[1]

Answer each question by circling the appropriate answer. Please respond to each item as honestly as possible; remember all of the information you provide is anonymous.

1. I am satisfied with my eating patterns.
    a. agree
    b. neutral
    c. disagree a little
    d. disagree
    e. disagree strongly

2. Would you presently call yourself a "binge eater"?
    a. yes, absolutely
    b. yes
    c. yes, probably
    d. yes, possibly
    e. no, probably not

3. Do you feel you have control over the amount of food you consume?
    a. most or all of the time
    b. a lot of the time
    c. occasionally
    d. rarely
    e. never

4. I am satisfied with the shape and size of my body.
    a. frequently or always
    b. sometimes
    c. occasionally
    d. rarely
    e. seldom or never

5. When I feel that my eating behavior is out of control, I try to take rather extreme measures to get back on course (strict dieting, fasting, laxatives, diuretics, self-induced vomiting, or vigorous exercise).
    a. always
    b. almost always
    c. frequently
    d. sometimes
    e. never or my eating behavior is never out of control

6. I am obsessed about the size and shape of my body.
    a. always
    b. almost always
    c. frequently
    d. sometimes
    e. seldom or never

7. There are times when I rapidly eat a very large amount of food.
    a. more than twice a week
    b. twice a week
    c. once a week
    d. 2-3 times a month
    e. once a month or less (or never)

---

[1]Revised with permission of M. Thelen.

8. How long have you been binge eating (eating uncontrollably to the point of stuffing yourself)?
    a. not applicable; I don't binge eat
    b. less than 3 months
    c. 3 months - 1 year
    d. 1 - 3 years
    e. 3 or more years

9. Most people I know would be amazed if they knew how much food I can consume at one sitting.
    a. without a doubt
    b. very probably
    c. probably
    d. possibly
    e. no

10. Compared with others your age, how preoccupied are you about your weight and body shape?
    a. a great deal more than average
    b. much more than average
    c. more than average
    d. a little more than average
    e. average or less than average

11. I am afraid to eat anything for fear that I won't be able to stop.
    a. always
    b. almost always
    c. frequently
    d. sometimes
    e. seldom or never

12. I feel tormented by the idea that I am fat or might gain weight.
    a. always
    b. almost always
    c. frequently
    d. sometimes
    e. seldom or never

13. How often do you intentionally vomit after eating?
    a. 2 or more times a week
    b. once a week
    c. 2-3 times a month
    d. once a month
    e. less than once a month or never

14. I eat a lot of food when I'm not even hungry.
    a. very frequently
    b. frequently
    c. occasionally
    d. sometimes
    e. seldom or never

15. My eating patterns are different from the eating patterns of most people.
    a. always
    b. almost always
    c. frequently
    d. sometimes
    e. seldom or never

16. After I binge eat I turn to one of several strict methods to try to keep from gaining weight (vigorous exercise, strict dieting, fasting, self-induced vomiting, laxatives, or diuretics).
    a. never or I don't binge eat
    b. rarely
    c. occasionally
    d. a lot of the time
    e. most or all of the time

17. When engaged in an eating binge, I tend to eat foods that are high in carbohydrates (sweets and starches).
    a. always
    b. almost always
    c. frequently
    d. sometimes
    e. seldom, or I don't binge

18. Compared to most people, my ability to control my eating behavior seems to be:
    a. greater than others' ability
    b. about the same
    c. less
    d. much less
    e. I have absolutely no control

19. I would presently label myself a 'compulsive eater' (one who engages in episodes of uncontrolled eating).
    a. absolutely
    b. yes
    c. yes, probably
    d. yes, possibly
    e. no, probably not

20. I hate the way my body looks after I eat too much.
    a. seldom or never
    b. sometimes
    c. frequently
    d. almost always
    e. always

21. When I am trying to keep from gaining weight, I feel that I have to resort to vigorous exercise, strict dieting, fasting, self-induced vomiting, laxatives, or diuretics.
    a. never
    b. rarely
    c. occasionally
    d. a lot of the time
    e. most of all of the time

22. Do you believe that it is easier for you to vomit than it is for most people?
    a. yes, it's no problem at all for me
    b. yes, it's easier
    c. yes, it's a little easier
    d. about the same
    e. no, it's less easy

23. I feel that food controls my life.
    a. always
    b. almost always
    c. frequently
    d. sometimes
    e. seldom or never

24. When consuming a large quantity of food, at what rate of speed do you usually eat?
    a. more rapidly than most people have ever eaten in their lives
    b. a lot more rapidly than most people
    c. a little more rapidly than most people
    d. about the same rate as most people
    e. more slowly than most people (or not applicable)

25. Right after I binge eat I feel:
    a. so fat and bloated I can't stand it
    b. extremely fat
    c. fat
    d. a little fat
    e. OK about how my body looks or I never binge eat

26. Compared to other people of my sex, my ability to always feel in control of how much I eat is:
    a. about the same or greater
    b. a little less
    c. less
    d. much less
    e. a great deal less

27. In the last 3 months, on the average how often did you binge eat (eat uncontrollably to the point of stuffing yourself)?
    a. once a month or less (or never)
    b. 2-3 times a month
    c. once a week
    d. twice a week
    e. more than twice a week

28. Most people I know would be surprised at how fat I look after I eat a lot of food.
    a. yes, definitely
    b. yes
    c. yes, probably
    d. yes, possibly
    e. no, probably not or I never eat a lot of food

29. What gender are you?
    a. male
    b. female

30. What age are you?
    a. Under 18
    b. 18 - 20
    c. 21 - 25
    d. 26 or older

THE BULIT-R[1]

Answer each question by circling the appropriate answer. Please respond to each item as honestly as possible; remember all of the information you provide is anonymous.

1. I am satisfied with my eating patterns.
   a. agree
   b. neutral
   c. disagree a little
   d. disagree
   e. disagree strongly

2. Would you presently call yourself a "binge eater"?
   a. yes, absolutely
   b. yes
   c. yes, probably
   d. yes, possibly
   e. no, probably not

3. Do you feel you have control over the amount of food you consume?
   a. most or all of the time
   b. a lot of the time
   c. occasionally
   d. rarely
   e. never

4. I am satisfied with the shape and size of my body.
   a. frequently or always
   b. sometimes
   c. occasionally
   d. rarely
   e. seldom or never

5. When I feel that my eating behavior is out of control, I try to take rather extreme measures to get back on course (strict dieting, fasting, laxatives, diuretics, self-induced vomiting, or vigorous exercise).
   a. always
   b. almost always
   c. frequently
   d. sometimes
   e. never or my eating behavior is never out of control

6. I am obsessed about the size and shape of my body.
   a. always
   b. almost always
   c. frequently
   d. sometimes
   e. seldom or never

7. There are times when I rapidly eat a very large amount of food.
   a. more than twice a week
   b. twice a week
   c. once a week
   d. 2-3 times a month
   e. once a month or less (or never)

---

[1]Revised with permission of M. Thelen.

8. How long have you been binge eating (eating uncontrollably to the point of stuffing yourself)?
   a. not applicable; I don't binge eat
   b. less than 3 months
   c. 3 months - 1 year
   d. 1 - 3 years
   e. 3 or more years

9. Most people I know would be amazed if they knew how much food I can consume at one sitting.
   a. without a doubt
   b. very probably
   c. probably
   d. possibly
   e. no

10. Compared with others your age, how preoccupied are you about your weight and body shape?
    a. a great deal more than average
    b. much more than average
    c. more than average
    d. a little more than average
    e. average or less than average

11. I am afraid to eat anything for fear that I won't be able to stop.
    a. always
    b. almost always
    c. frequently
    d. sometimes
    e. seldom or never

12. I feel tormented by the idea that I am fat or might gain weight.
    a. always
    b. almost always
    c. frequently
    d. sometimes
    e. seldom or never

13. How often do you intentionally vomit after eating?
    a. 2 or more times a week
    b. once a week
    c. 2-3 times a month
    d. once a month
    e. less than once a month or never

14. I eat a lot of food when I'm not even hungry.
    a. very frequently
    b. frequently
    c. occasionally
    d. sometimes
    e. seldom or never

15. My eating patterns are different from the eating patterns of most people.
    a. always
    b. almost always
    c. frequently
    d. sometimes
    e. seldom or never

16. After I binge eat I turn to one of several strict methods to try to keep from gaining weight (vigorous exercise, strict dieting, fasting, self-induced vomiting, laxatives, or diuretics).
    a. never or I don't binge eat
    b. rarely
    c. occasionally
    d. a lot of the time
    e. most or all of the time

17. When engaged in an eating binge, I tend to eat foods that are high in carbohydrates (sweets and starches).
    a. always
    b. almost always
    c. frequently
    d. sometimes
    e. seldom, or I don't binge

18. Compared to most people, my ability to control my eating behavior seems to be:
    a. greater than others' ability
    b. about the same
    c. less
    d. much less
    e. I have absolutely no control

19. I would presently label myself a 'compulsive eater' (one who engages in episodes of uncontrolled eating).
    a. absolutely
    b. yes
    c. yes, probably
    d. yes, possibly
    e. no, probably not

20. I hate the way my body looks after I eat too much.
    a. seldom or never
    b. sometimes
    c. frequently
    d. almost always
    e. always

21. When I am trying to keep from gaining weight, I feel that I have to resort to vigorous exercise, strict dieting, fasting, self-induced vomiting, laxatives, or diuretics.
    a. never
    b. rarely
    c. occasionally
    d. a lot of the time
    e. most of all of the time

22. Do you believe that it is easier for you to vomit than it is for most people?
    a. yes, it's no problem at all for me
    b. yes, it's easier
    c. yes, it's a little easier
    d. about the same
    e. no, it's less easy

23. I feel that food controls my life.
    a. always
    b. almost always
    c. frequently
    d. sometimes
    e. seldom or never

24. When consuming a large quantity of food, at what rate of speed do you usually eat?
    a. more rapidly than most people have ever eaten in their lives
    b. a lot more rapidly than most people
    c. a little more rapidly than most people
    d. about the same rate as most people
    e. more slowly than most people (or not applicable)

25. Right after I binge eat I feel:
    a. so fat and bloated I can't stand it
    b. extremely fat
    c. fat
    d. a little fat
    e. OK about how my body looks or I never binge eat

26. Compared to other people of my sex, my ability to always feel in control of how much I eat is:
    a. about the same or greater
    b. a little less
    c. less
    d. much less
    e. a great deal less

27. In the last 3 months, on the average how often did you binge eat (eat uncontrollably to the point of stuffing yourself)?
    a. once a month or less (or never)
    b. 2-3 times a month
    c. once a week
    d. twice a week
    e. more than twice a week

28. Most people I know would be surprised at how fat I look after I eat a lot of food.
    a. yes, definitely
    b. yes
    c. yes, probably
    d. yes, possibly
    e. no, probably not or I never eat a lot of food

29. What gender are you?
    a. male
    b. female

30. What age are you?
    a. Under 18
    b. 18 - 20
    c. 21 - 25
    d. 26 or older

THE BULIT-R[1]

Answer each question by circling the appropriate answer. Please respond to each item as honestly as possible; remember all of the information you provide is anonymous.

1. I am satisfied with my eating patterns.
   a. agree
   b. neutral
   c. disagree a little
   d. disagree
   e. disagree strongly

2. Would you presently call yourself a "binge eater"?
   a. yes, absolutely
   b. yes
   c. yes, probably
   d. yes, possibly
   e. no, probably not

3. Do you feel you have control over the amount of food you consume?
   a. most or all of the time
   b. a lot of the time
   c. occasionally
   d. rarely
   e. never

4. I am satisfied with the shape and size of my body.
   a. frequently or always
   b. sometimes
   c. occasionally
   d. rarely
   e. seldom or never

5. When I feel that my eating behavior is out of control, I try to take rather extreme measures to get back on course (strict dieting, fasting, laxatives, diuretics, self-induced vomiting, or vigorous exercise).
   a. always
   b. almost always
   c. frequently
   d. sometimes
   e. never or my eating behavior is never out of control

6. I am obsessed about the size and shape of my body.
   a. always
   b. almost always
   c. frequently
   d. sometimes
   e. seldom or never

7. There are times when I rapidly eat a very large amount of food.
   a. more than twice a week
   b. twice a week
   c. once a week
   d. 2-3 times a month
   e. once a month or less (or never)

---

[1]Revised with permission of M. Thelen.

8. How long have you been binge eating (eating uncontrollably to the point of stuffing yourself)?
   a. not applicable; I don't binge eat
   b. less than 3 months
   c. 3 months - 1 year
   d. 1 - 3 years
   e. 3 or more years

9. Most people I know would be amazed if they knew how much food I can consume at one sitting.
   a. without a doubt
   b. very probably
   c. probably
   d. possibly
   e. no

10. Compared with others your age, how preoccupied are you about your weight and body shape?
    a. a great deal more than average
    b. much more than average
    c. more than average
    d. a little more than average
    e. average or less than average

11. I am afraid to eat anything for fear that I won't be able to stop.
    a. always
    b. almost always
    c. frequently
    d. sometimes
    e. seldom or never

12. I feel tormented by the idea that I am fat or might gain weight.
    a. always
    b. almost always
    c. frequently
    d. sometimes
    e. seldom or never

13. How often do you intentionally vomit after eating?
    a. 2 or more times a week
    b. once a week
    c. 2-3 times a month
    d. once a month
    e. less than once a month or never

14. I eat a lot of food when I'm not even hungry.
    a. very frequently
    b. frequently
    c. occasionally
    d. sometimes
    e. seldom or never

15. My eating patterns are different from the eating patterns of most people.
    a. always
    b. almost always
    c. frequently
    d. sometimes
    e. seldom or never

16. After I binge eat I turn to one of several strict methods to try to keep from gaining weight (vigorous exercise, strict dieting, fasting, self-induced vomiting, laxatives, or diuretics).
    a. never or I don't binge eat
    b. rarely
    c. occasionally
    d. a lot of the time
    e. most or all of the time

17. When engaged in an eating binge, I tend to eat foods that are high in carbohydrates (sweets and starches).
    a. always
    b. almost always
    c. frequently
    d. sometimes
    e. seldom, or I don't binge

18. Compared to most people, my ability to control my eating behavior seems to be:
    a. greater than others' ability
    b. about the same
    c. less
    d. much less
    e. I have absolutely no control

19. I would presently label myself a 'compulsive eater' (one who engages in episodes of uncontrolled eating).
    a. absolutely
    b. yes
    c. yes, probably
    d. yes, possibly
    e. no, probably not

20. I hate the way my body looks after I eat too much.
    a. seldom or never
    b. sometimes
    c. frequently
    d. almost always
    e. always

21. When I am trying to keep from gaining weight, I feel that I have to resort to vigorous exercise, strict dieting, fasting, self-induced vomiting, laxatives, or diuretics.
    a. never
    b. rarely
    c. occasionally
    d. a lot of the time
    e. most of all of the time

22. Do you believe that it is easier for you to vomit than it is for most people?
    a. yes, it's no problem at all for me
    b. yes, it's easier
    c. yes, it's a little easier
    d. about the same
    e. no, it's less easy

23. I feel that food controls my life.
    a. always
    b. almost always
    c. frequently
    d. sometimes
    e. seldom or never

24. When consuming a large quantity of food, at what rate of speed do you usually eat?
    a. more rapidly than most people have ever eaten in their lives
    b. a lot more rapidly than most people
    c. a little more rapidly than most people
    d. about the same rate as most people
    e. more slowly than most people (or not applicable)

25. Right after I binge eat I feel:
    a. so fat and bloated I can't stand it
    b. extremely fat
    c. fat
    d. a little fat
    e. OK about how my body looks or I never binge eat

26. Compared to other people of my sex, my ability to always feel in control of how much I eat is:
    a. about the same or greater
    b. a little less
    c. less
    d. much less
    e. a great deal less

27. In the last 3 months, on the average how often did you binge eat (eat uncontrollably to the point of stuffing yourself)?
    a. once a month or less (or never)
    b. 2-3 times a month
    c. once a week
    d. twice a week
    e. more than twice a week

28. Most people I know would be surprised at how fat I look after I eat a lot of food.
    a. yes, definitely
    b. yes
    c. yes, probably
    d. yes, possibly
    e. no, probably not or I never eat a lot of food

29. What gender are you?
    a. male
    b. female

30. What age are you?
    a. Under 18
    b. 18 - 20
    c. 21 - 25
    d. 26 or older

THE BULIT-R[1]

Answer each question by circling the appropriate answer. Please respond to each item as honestly as possible; remember all of the information you provide is anonymous.

1. I am satisfied with my eating patterns.
   a. agree
   b. neutral
   c. disagree a little
   d. disagree
   e. disagree strongly

2. Would you presently call yourself a "binge eater"?
   a. yes, absolutely
   b. yes
   c. yes, probably
   d. yes, possibly
   e. no, probably not

3. Do you feel you have control over the amount of food you consume?
   a. most or all of the time
   b. a lot of the time
   c. occasionally
   d. rarely
   e. never

4. I am satisfied with the shape and size of my body.
   a. frequently or always
   b. sometimes
   c. occasionally
   d. rarely
   e. seldom or never

5. When I feel that my eating behavior is out of control, I try to take rather extreme measures to get back on course (strict dieting, fasting, laxatives, diuretics, self-induced vomiting, or vigorous exercise).
   a. always
   b. almost always
   c. frequently
   d. sometimes
   e. never or my eating behavior is never out of control

6. I am obsessed about the size and shape of my body.
   a. always
   b. almost always
   c. frequently
   d. sometimes
   e. seldom or never

7. There are times when I rapidly eat a very large amount of food.
   a. more than twice a week
   b. twice a week
   c. once a week
   d. 2-3 times a month
   e. once a month or less (or never)

---

[1]Revised with permission of M. Thelen.

8. How long have you been binge eating (eating uncontrollably to the point of stuffing yourself)?
   a. not applicable; I don't binge eat
   b. less than 3 months
   c. 3 months - 1 year
   d. 1 - 3 years
   e. 3 or more years

9. Most people I know would be amazed if they knew how much food I can consume at one sitting.
   a. without a doubt
   b. very probably
   c. probably
   d. possibly
   e. no

10. Compared with others your age, how preoccupied are you about your weight and body shape?
    a. a great deal more than average
    b. much more than average
    c. more than average
    d. a little more than average
    e. average or less than average

11. I am afraid to eat anything for fear that I won't be able to stop.
    a. always
    b. almost always
    c. frequently
    d. sometimes
    e. seldom or never

12. I feel tormented by the idea that I am fat or might gain weight.
    a. always
    b. almost always
    c. frequently
    d. sometimes
    e. seldom or never

13. How often do you intentionally vomit after eating?
    a. 2 or more times a week
    b. once a week
    c. 2-3 times a month
    d. once a month
    e. less than once a month or never

14. I eat a lot of food when I'm not even hungry.
    a. very frequently
    b. frequently
    c. occasionally
    d. sometimes
    e. seldom or never

15. My eating patterns are different from the eating patterns of most people.
    a. always
    b. almost always
    c. frequently
    d. sometimes
    e. seldom or never

16. After I binge eat I turn to one of several strict methods to try to keep from gaining weight (vigorous exercise, strict dieting, fasting, self-induced vomiting, laxatives, or diuretics).
    a. never or I don't binge eat
    b. rarely
    c. occasionally
    d. a lot of the time
    e. most or all of the time

17. When engaged in an eating binge, I tend to eat foods that are high in carbohydrates (sweets and starches).
    a. always
    b. almost always
    c. frequently
    d. sometimes
    e. seldom, or I don't binge

18. Compared to most people, my ability to control my eating behavior seems to be:
    a. greater than others' ability
    b. about the same
    c. less
    d. much less
    e. I have absolutely no control

19. I would presently label myself a 'compulsive eater' (one who engages in episodes of uncontrolled eating).
    a. absolutely
    b. yes
    c. yes, probably
    d. yes, possibly
    e. no, probably not

20. I hate the way my body looks after I eat too much.
    a. seldom or never
    b. sometimes
    c. frequently
    d. almost always
    e. always

21. When I am trying to keep from gaining weight, I feel that I have to resort to vigorous exercise, strict dieting, fasting, self-induced vomiting, laxatives, or diuretics.
    a. never
    b. rarely
    c. occasionally
    d. a lot of the time
    e. most of all of the time

22. Do you believe that it is easier for you to vomit than it is for most people?
    a. yes, it's no problem at all for me
    b. yes, it's easier
    c. yes, it's a little easier
    d. about the same
    e. no, it's less easy

23. I feel that food controls my life.
    a. always
    b. almost always
    c. frequently
    d. sometimes
    e. seldom or never

24. When consuming a large quantity of food, at what rate of speed do you usually eat?
    a. more rapidly than most people have ever eaten in their lives
    b. a lot more rapidly than most people
    c. a little more rapidly than most people
    d. about the same rate as most people
    e. more slowly than most people (or not applicable)

25. Right after I binge eat I feel:
    a. so fat and bloated I can't stand it
    b. extremely fat
    c. fat
    d. a little fat
    e. OK about how my body looks or I never binge eat

26. Compared to other people of my sex, my ability to always feel in control of how much I eat is:
    a. about the same or greater
    b. a little less
    c. less
    d. much less
    e. a great deal less

27. In the last 3 months, on the average how often did you binge eat (eat uncontrollably to the point of stuffing yourself)?
    a. once a month or less (or never)
    b. 2-3 times a month
    c. once a week
    d. twice a week
    e. more than twice a week

28. Most people I know would be surprised at how fat I look after I eat a lot of food.
    a. yes, definitely
    b. yes
    c. yes, probably
    d. yes, possibly
    e. no, probably not or I never eat a lot of food

29. What gender are you?
    a. male
    b. female

30. What age are you?
    a. Under 18
    b. 18 - 20
    c. 21 - 25
    d. 26 or older

EXTRASENSORY PERCEPTION (ESP) SURVEY

This is an anonymous survey, conducted as a project for a course in research methods, about your attitudes toward extrasensory perception (ESP) and whether you have had any ESP experiences.

Please check "yes" or "no" to each of the following statements. If you are unsure, check "no". Do not leave any blank.

		Yes	No
1.	I believe in the existence of ESP.	___	___
2.	I believe I have had at least one ESP experience.	___	___
3.	I believe ghosts exist.	___	___
4.	I believe in life after death.	___	___
5.	I believe that some people can contact people who have died.	___	___
6.	I believe that there are flying saucers and people from other places than earth visiting our planet.	___	___
7.	I have had a telepathic experience, where I felt like I was reading another person's thoughts.	___	___
8.	I have had a specific dream about something which matched in detail an event which occurred after my dream. I did not know about the event at the time of the dream and did not expect it.	___	___
9.	I have had the experience of feeling that "I" was outside of or away from my body. This is called an out-of-body experience or astral projection.	___	___
10.	I have seen, heard, or been touched by another being, often referred to as a ghost, and could not explain the experience as being due to a physical or natural cause.	___	___
11.	I have moved an object with my thoughts alone. This is called psychokinesis (PK).	___	___
12.	I have seen light around the body or body parts of another person. As far as I could tell this was not due to a physical or natural cause. This experience is called an aura.	___	___

Demographic Information:

AGE: _____   GENDER: MALE: _____   FEMALE: _____

EXTRASENSORY PERCEPTION (ESP) SURVEY

This is an anonymous survey, conducted as a project for a course in research methods, about your attitudes toward extrasensory perception (ESP) and whether you have had any ESP experiences.

Please check "yes" or "no" to each of the following statements. If you are unsure, check "no". Do not leave any blank.

		Yes	No
1.	I believe in the existence of ESP.	___	___
2.	I believe I have had at least one ESP experience.	___	___
3.	I believe ghosts exist.	___	___
4.	I believe in life after death.	___	___
5.	I believe that some people can contact people who have died.	___	___
6.	I believe that there are flying saucers and people from other places than earth visiting our planet.	___	___
7.	I have had a telepathic experience, where I felt like I was reading another person's thoughts.	___	___
8.	I have had a specific dream about something which matched in detail an event which occurred after my dream. I did not know about the event at the time of the dream and did not expect it.	___	___
9.	I have had the experience of feeling that "I" was outside of or away from my body. This is called an out-of-body experience or astral projection.	___	___
10.	I have seen, heard, or been touched by another being, often referred to as a ghost, and could not explain the experience as being due to a physical or natural cause.	___	___
11.	I have moved an object with my thoughts alone. This is called psychokinesis (PK).	___	___
12.	I have seen light around the body or body parts of another person. As far as I could tell this was not due to a physical or natural cause. This experience is called an aura.	___	___

Demographic Information:

AGE: _____   GENDER: MALE: _____   FEMALE: _____

EXTRASENSORY PERCEPTION (ESP) SURVEY

This is an anonymous survey, conducted as a project for a course in research methods, about your attitudes toward extrasensory perception (ESP) and whether you have had any ESP experiences.

Please check "yes" or "no" to each of the following statements. If you are unsure, check "no". Do not leave any blank.

		Yes	No
1.	I believe in the existence of ESP.	___	___
2.	I believe I have had at least one ESP experience.	___	___
3.	I believe ghosts exist.	___	___
4.	I believe in life after death.	___	___
5.	I believe that some people can contact people who have died.	___	___
6.	I believe that there are flying saucers and people from other places than earth visiting our planet.	___	___
7.	I have had a telepathic experience, where I felt like I was reading another person's thoughts.	___	___
8.	I have had a specific dream about something which matched in detail an event which occurred after my dream. I did not know about the event at the time of the dream and did not expect it.	___	___
9.	I have had the experience of feeling that "I" was outside of or away from my body. This is called an out-of-body experience or astral projection.	___	___
10.	I have seen, heard, or been touched by another being, often referred to as a ghost, and could not explain the experience as being due to a physical or natural cause.	___	___
11.	I have moved an object with my thoughts alone. This is called psychokinesis (PK).	___	___
12.	I have seen light around the body or body parts of another person. As far as I could tell this was not due to a physical or natural cause. This experience is called an aura.	___	___

Demographic Information:

AGE: _____   GENDER: MALE: _____ FEMALE: _____

EXTRASENSORY PERCEPTION (ESP) SURVEY

This is an anonymous survey, conducted as a project for a course in research methods, about your attitudes toward extrasensory perception (ESP) and whether you have had any ESP experiences.

Please check "yes" or "no" to each of the following statements. If you are unsure, check "no". Do not leave any blank.

	Yes	No
1. I believe in the existence of ESP.	___	___
2. I believe I have had at least one ESP experience.	___	___
3. I believe ghosts exist.	___	___
4. I believe in life after death.	___	___
5. I believe that some people can contact people who have died.	___	___
6. I believe that there are flying saucers and people from other places than earth visiting our planet.	___	___
7. I have had a telepathic experience, where I felt like I was reading another person's thoughts.	___	___
8. I have had a specific dream about something which matched in detail an event which occurred after my dream. I did not know about the event at the time of the dream and did not expect it.	___	___
9. I have had the experience of feeling that "I" was outside of or away from my body. This is called an out-of-body experience or astral projection.	___	___
10. I have seen, heard, or been touched by another being, often referred to as a ghost, and could not explain the experience as being due to a physical or natural cause.	___	___
11. I have moved an object with my thoughts alone. This is called psychokinesis (PK).	___	___
12. I have seen light around the body or body parts of another person. As far as I could tell this was not due to a physical or natural cause. This experience is called an aura.	___	___

Demographic Information:

AGE: _____   GENDER: MALE: _____   FEMALE: _____

EXTRASENSORY PERCEPTION (ESP) SURVEY

This is an anonymous survey, conducted as a project for a course in research methods, about your attitudes toward extrasensory perception (ESP) and whether you have had any ESP experiences.

Please check "yes" or "no" to each of the following statements. If you are unsure, check "no". Do not leave any blank.

		Yes	No
1.	I believe in the existence of ESP.	—	—
2.	I believe I have had at least one ESP experience.	—	—
3.	I believe ghosts exist.	—	—
4.	I believe in life after death.	—	—
5.	I believe that some people can contact people who have died.	—	—
6.	I believe that there are flying saucers and people from other places than earth visiting our planet.	—	—
7.	I have had a telepathic experience, where I felt like I was reading another person's thoughts.	—	—
8.	I have had a specific dream about something which matched in detail an event which occurred after my dream. I did not know about the event at the time of the dream and did not expect it.	—	—
9.	I have had the experience of feeling that "I" was outside of or away from my body. This is called an out-of-body experience or astral projection.	—	—
10.	I have seen, heard, or been touched by another being, often referred to as a ghost, and could not explain the experience as being due to a physical or natural cause.	—	—
11.	I have moved an object with my thoughts alone. This is called psychokinesis (PK).	—	—
12.	I have seen light around the body or body parts of another person. As far as I could tell this was not due to a physical or natural cause. This experience is called an aura.	—	—

Demographic Information:

AGE: _____   GENDER: MALE: _____   FEMALE: _____

APPENDIX B

Materials for Topic 7: Independent and Dependent Variables

VILLAGE CORNER	MAGAZINE BOULDER	UNIVERSITY MAIDEN
OFFICER GIRL	HAMMER ELEPHANT	BUILDING CIRCLE
DOLLAR TEMPLE	FURNITURE BODY	CLOCK ENGINE
GRANDMOTHER ELBOW	FISHERMAN PRISONER	HOTEL LETTER
GARDEN ORCHESTRA	ROCK HOUSE	DIAMOND GENTLEMAN

High Imagery Pair

High Imagery Pair

High Imagery Pair

High Imagery Pair

High Imagery Pair

High Imagery Pair

HIgh Imagery Pair

High Imagery Pair

High Imagery Pair

High Imagery Pair

High Imagery Pair

High Imagery Pair

High Imagery Pair

High Imagery Pair

High Imagery Pair

FATE MEMORY	ANSWER MOMENT	CONFIDENCE MIND
EVIDENCE SENTIMENT	KNOWLEDGE SPIRIT	THOUGHT UNIT
AGREEMENT ABILITY	ATTITUDE CHANCE	PATENT INTELLECT
SITUATION THEORY	PERMISSION HONOR	HISTORY EXPRESSION
SOUL FACT	NECESSITY BELIEF	QUALITY CUSTOM

Low Imagery Pair	**Low Imagery Pair**	**Low Imagery Pair**
Low Imagery Pair	**Low Imagery Pair**	**Low Imagery Pair**
Low Imagery Pair	**Low Imagery Pair**	**Low Imagery Pair**
Low Imagery Pair	**Low Imagery Pair**	**Low Imagery Pair**
Low Imagery Pair	**Low Imagery Pair**	**Low Imagery Pair**

ENQUIRY PANEL	SYRUP TRICK	CONTINUATION SHAME
BLOUSE PANSY	SNOW ORANGE	SPARROW ANNIVERSARY
DEBATE SHINGLE	KITE SNEEZE	SPARK CONTRAST
SHED CENSURE		

Filler

Filler

Filler

Filler

Filler

Filler

Filler

Filler

Filler

Filler

UNIVERSITY	FISHERMAN	CIRCLE
BODY	DOLLAR	VILLAGE
ELBOW	DIAMOND	CLOCK
MAGAZINE	HAMMER	LETTER
GARDEN	OFFICER	HOUSE

High Response   High Response   High Response

High Response   High Response   High Response

High Response   High Response   High Response

High Response   High Response   High Response

High Response   High Response   High Response

KNOWLEDGE	AGREEMENT	SOUL
CONFIDENCE	FATE	PERMISSION
PATENT	UNIT	EXPRESSION
BELIEF	CHANCE	SENTIMENT
MOMENT	THEORY	QUALITY

Low Respose

Low Response

Low Response

Low Response

Low Response

Low Response

Low Response

Low Response

Low Response

Low Response

Low Response

Low Response

Low Response

Low Response

Low Response

This is an anonymous questionnaire to assess the manner in which people judge various offenses. Below is a brief account of a criminal offense. When you have finished reading the case account, you will be asked to give your personal opinion concerning the case. That is, you are to sentence the defendant described in the case account to a specific number of years of imprisonment. Take as much time as you want in reading and contemplating the case before you. Finally, sentence the defendant. Remember that we are interested in your personal opinion, so please give your own personal judgment and not how you feel others might react to the case or how you feel you should react to it. One other thing -- in making your sentence, consider the question of parole as being beyond your jurisdiction. That is, sentence the defendant irrespective of whether or not you feel he should have opportunity for parole after a certain number of years in prison.

John Sander was driving home from an annual Christmas office party on the evening of December 24 when his automobile struck and killed a pedestrian by the name of Martin Lowe. The circumstances leading to this event were as follows: The employees of the insurance office where Sander worked began to party at around 2:00 P.M. on the afternoon of the 24th. By 5:00 P.M. some people were already leaving for home, although many continued to drink and socialize. Sander, who by this time had had several drinks, was offered a lift home by a friend who did not drink and who suggested that Sander leave his car at the office and pick it up when he was in "better shape." Sander declined the offer, claiming he was "stone sober" and would manage fine. By the time Sander had finished another drink, the party was beginning to break up. Sander left the office building and walked to the garage where he had parked his car, a four-door 1965 Chevrolet. It had just started to snow. He wished the garage attendant a Merry Christmas and pulled out into the street. Traffic was very heavy at the time. Sander was six blocks from the garage when he was stopped by a policeman for reckless driving. It was quite apparent to the officer that Sander had been drinking, but rather than give him a ticket on Christmas Eve, he said that he would let Sander off if he would promise to leave his car and take a taxi. Sander agreed. The officer hailed a taxi and Sander got into it. The minute the taxi had turned a corner, however, Sander told the driver to pull over to the curb and let him out. Sander paid the driver and started back to where he had parked his own car. Upon reaching his car he proceeded to start it up and drove off.

He had driven four blocks from the street where the police officer had stopped him when he ran a red light and stuck Lowe, who was crossing the street. Sander immediately stopped the car. Lowe died a few minutes later on the way to the hospital. It was later ascertained that internal hemorrhaging was the cause of death. Sander was apprehended and charged with negligent homicide. The police medical examiner's report indicated that Sander's estimated blood alcohol concentration was between 2.5 and 3.0% at the time of the accident.

Lowe is a noted architect and prominent member of the community. He had designed many well-known buildings throughout the state . . . was an active member of the community welfare board. At the time of the incident, Lowe was on his way to the Lincoln Orphanage, of which he was a founding member, with Christmas gifts. He is survived by his wife and two children, ages 11 and 15. Sander is a sixty-four-year-old insurance adjustor who has been employed by the same insurance firm for 42 years. Sander was friendly with everyone and was known as a good worker. Sander is a widower, his wife having died of cancer the previous year, and he is, consequently, spending Christmas Eve with his son and daughter-in-law. When the incident occurred, Sander's leg banged the steering column, reaggravating a gun wound which had been the source of a slight limp and much pain. Sander's traffic record shows he has received three tickets in the past five years, two of which were moving violations.

Sander was charged with negligent automobile homicide, a crime which in the state is punishable by imprisonment of one to twenty-five years.

Your job is to determine how many years of imprisonment, according to your own judgement, he should receive. After you have considered the case presented on the previous page, note below how many years of imprisonment you, as judge, would give him. Parole is beyond your jurisdiction. That is, sentence the defendant irrespective of whether or not you feel he should have opportunity for parole after a certain number of years in prison.

NUMBER OF YEARS _____

Now, would you indicate your impression of the defendant and the victim on the following scale. Circle the number which is closest to your impression of each person.

SANDERS;

extremely									extremely
	1	2	3	4	5	6	7	8	9
favorable									unfavorable

LOWE:

extremely									extremely
	1	2	3	4	5	6	7	8	9
favorable									unfavorable

What issues did you think were important in deciding the length of imprisonment?

What is your age? _____   Gender: Male _____   Female _____

Thank you very much for your participation. Please put this in the envelope the interviewer has so that your questionnaire will remain completely anonymous.

(1)

This is an anonymous questionnaire to assess the manner in which people judge various offenses. Below is a brief account of a criminal offense. When you have finished reading the case account, you will be asked to give your personal opinion concerning the case. That is, you are to sentence the defendant described in the case account to a specific number of years of imprisonment. Take as much time as you want in reading and contemplating the case before you. Finally, sentence the defendant. Remember that we are interested in your personal opinion, so please give your own personal judgment and not how you feel others might react to the case or how you feel you should react to it. One other thing -- in making your sentence, consider the question of parole as being beyond your jurisdiction. That is, sentence the defendant irrespective of whether or not you feel he should have opportunity for parole after a certain number of years in prison.

John Sander was driving home from an annual Christmas office party on the evening of December 24 when his automobile struck and killed a pedestrian by the name of Martin Lowe. The circumstances leading to this event were as follows: The employees of the insurance office where Sander worked began to party at around 2:00 P.M. on the afternoon of the 24th. By 5:00 P.M. some people were already leaving for home, although many continued to drink and socialize. Sander, who by this time had had several drinks, was offered a lift home by a friend who did not drink and who suggested that Sander leave his car at the office and pick it up when he was in "better shape." Sander declined the offer, claiming he was "stone sober" and would manage fine. By the time Sander had finished another drink, the party was beginning to break up. Sander left the office building and walked to the garage where he had parked his car, a four-door 1965 Chevrolet. It had just started to snow. He wished the garage attendant a Merry Christmas and pulled out into the street. Traffic was very heavy at the time. Sander was six blocks from the garage when he was stopped by a policeman for reckless driving. It was quite apparent to the officer that Sander had been drinking, but rather than give him a ticket on Christmas Eve, he said that he would let Sander off if he would promise to leave his car and take a taxi. Sander agreed. The officer hailed a taxi and Sander got into it. The minute the taxi had turned a corner, however, Sander told the driver to pull over to the curb and let him out. Sander paid the driver and started back to where he had parked his own car. Upon reaching his car he proceeded to start it up and drove off.

He had driven four blocks from the street where the police officer had stopped him when he ran a red light and stuck Lowe, who was crossing the street. Sander immediately stopped the car. Lowe died a few minutes later on the way to the hospital. It was later ascertained that internal hemorrhaging was the cause of death. Sander was apprehended and charged with negligent homicide. The police medical examiner's report indicated that Sander's estimated blood alcohol concentration was between 2.5 and 3.0% at the time of the accident.

Lowe is a notorious gangster and syndicate boss who had been vying for power in the syndicate controlling the state's underworld activities. He was best known for his alleged responsibility in the Riverview massacre of five men. At the time of the incident, Lowe was carrying a loaded 32-caliber pistol which was found on his body. He had been out of jail on bond, awaiting trial on a double indictment of mail fraud and income tax evasion. Sander is a sixty-four-year-old insurance adjustor who has been employed by the same insurance firm for 42 years. Sander was friendly with everyone and was known as a good worker. Sander is a widower, his wife having died of cancer the previous year, and he is, consequently, spending Christmas Eve with his son and daughter-in-law. When the incident occurred, Sander's leg banged the steering column, reaggravating a gun wound which had been the source of a slight limp and much pain. Sander's traffic record shows he has received three tickets in the past five years, two of which were moving violations.

Sander was charged with negligent automobile homicide, a crime which in the state is punishable by imprisonment of one to twenty-five years.

Your job is to determine how many years of imprisonment, according to your own judgement, he should receive. After you have considered the case presented on the previous page, note below how many years of imprisonment you, as judge, would give him. Parole is beyond your jurisdiction. That is, sentence the defendant irrespective of whether or not you feel he should have opportunity for parole after a certain number of years in prison.

NUMBER OF YEARS _____

Now, would you indicate your impression of the defendant and the victim on the following scale. Circle the number which is closest to your impression of each person.

SANDERS;

extremely									extremely
	1	2	3	4	5	6	7	8	9
favorable									unfavorable

LOWE:

extremely									extremely
	1	2	3	4	5	6	7	8	9
favorable									unfavorable

What issues did you think were important in deciding the length of imprisonment?

What is your age? _____   Gender: Male _____   Female _____

Thank you very much for your participation. Please put this in the envelope the interviewer has so that your questionnaire will remain completely anonymous.

This is an anonymous questionnaire to assess the manner in which people judge various offenses. Below is a brief account of a criminal offense. When you have finished reading the case account, you will be asked to give your personal opinion concerning the case. That is, you are to sentence the defendant described in the case account to a specific number of years of imprisonment. Take as much time as you want in reading and contemplating the case before you. Finally, sentence the defendant. Remember that we are interested in your personal opinion, so please give your own personal judgment and not how you feel others might react to the case or how you feel you should react to it. One other thing -- in making your sentence, consider the question of parole as being beyond your jurisdiction. That is, sentence the defendant irrespective of whether or not you feel he should have opportunity for parole after a certain number of years in prison.

John Sander was driving home from an annual Christmas office party on the evening of December 24 when his automobile struck and killed a pedestrian by the name of Martin Lowe. The circumstances leading to this event were as follows: The employees of the insurance office where Sander worked began to party at around 2:00 P.M. on the afternoon of the 24th. By 5:00 P.M. some people were already leaving for home, although many continued to drink and socialize. Sander, who by this time had had several drinks, was offered a lift home by a friend who did not drink and who suggested that Sander leave his car at the office and pick it up when he was in "better shape." Sander declined the offer, claiming he was "stone sober" and would manage fine. By the time Sander had finished another drink, the party was beginning to break up. Sander left the office building and walked to the garage where he had parked his car, a four-door 1965 Chevrolet. It had just started to snow. He wished the garage attendant a Merry Christmas and pulled out into the street. Traffic was very heavy at the time. Sander was six blocks from the garage when he was stopped by a policeman for reckless driving. It was quite apparent to the officer that Sander had been drinking, but rather than give him a ticket on Christmas Eve, he said that he would let Sander off if he would promise to leave his car and take a taxi. Sander agreed. The officer hailed a taxi and Sander got into it. The minute the taxi had turned a corner, however, Sander told the driver to pull over to the curb and let him out. Sander paid the driver and started back to where he had parked his own car. Upon reaching his car he proceeded to start it up and drove off.

He had driven four blocks from the street where the police officer had stopped him when he ran a red light and stuck Lowe, who was crossing the street. Sander immediately stopped the car. Lowe died a few minutes later on the way to the hospital. It was later ascertained that internal hemorrhaging was the cause of death. Sander was apprehended and charged with negligent homicide. The police medical examiner's report indicated that Sander's estimated blood alcohol concentration was between 2.5 and 3.0% at the time of the accident.

Lowe is a noted architect and prominent member of the community. He had designed many well-known buildings throughout the state . . . was an active member of the community welfare board. At the time of the incident, Lowe was on his way to the Lincoln Orphanage, of which he was a founding member, with Christmas gifts. He is survived by his wife and two children, ages 11 and 15. Sander is a thirty-three-year-old janitor. In the building where Sander has been working as a janitor for the past two months, he was not known by many of the firm employees, but was nevertheless invited to join the party. Sander is a two-time divorcee, with three children by his first wife, who has since remarried. He was going to spend Christmas Eve with his girlfriend in her apartment. The effect of the incident on Sander was negligible; he was slightly shaken up by the impact, but suffered no major injuries. Sander has two misdemeanors on his criminal record in the past five years -- breaking and entering and a drug violation. His traffic record shows three tickets in the same space of time.

Sander was charged with negligent automobile homicide, a crime which in the state is punishable by imprisonment of one to twenty-five years.

Your job is to determine how many years of imprisonment, according to your own judgement, he should receive. After you have considered the case presented on the previous page, note below how many years of imprisonment you, as judge, would give him. Parole is beyond your jurisdiction. That is, sentence the defendant irrespective of whether or not you feel he should have opportunity for parole after a certain number of years in prison.

NUMBER OF YEARS _____

Now, would you indicate your impression of the defendant and the victim on the following scale. Circle the number which is closest to your impression of each person.

SANDERS;

extremely									extremely
	1	2	3	4	5	6	7	8	9
favorable									unfavorable

LOWE:

extremely									extremely
	1	2	3	4	5	6	7	8	9
favorable									unfavorable

What issues did you think were important in deciding the length of imprisonment?

What is your age? _____   Gender: Male _____   Female _____

Thank you very much for your participation. Please put this in the envelope the interviewer has so that your questionnaire will remain completely anonymous.

This is an anonymous questionnaire to assess the manner in which people judge various offenses. Below is a brief account of a criminal offense. When you have finished reading the case account, you will be asked to give your personal opinion concerning the case. That is, you are to sentence the defendant described in the case account to a specific number of years of imprisonment. Take as much time as you want in reading and contemplating the case before you. Finally, sentence the defendant. Remember that we are interested in your personal opinion, so please give your own personal judgment and not how you feel others might react to the case or how you feel you should react to it. One other thing -- in making your sentence, consider the question of parole as being beyond your jurisdiction. That is, sentence the defendant irrespective of whether or not you feel he should have opportunity for parole after a certain number of years in prison.

John Sander was driving home from an annual Christmas office party on the evening of December 24 when his automobile struck and killed a pedestrian by the name of Martin Lowe. The circumstances leading to this event were as follows: The employees of the insurance office where Sander worked began to party at around 2:00 P.M. on the afternoon of the 24th. By 5:00 P.M. some people were already leaving for home, although many continued to drink and socialize. Sander, who by this time had had several drinks, was offered a lift home by a friend who did not drink and who suggested that Sander leave his car at the office and pick it up when he was in "better shape." Sander declined the offer, claiming he was "stone sober" and would manage fine. By the time Sander had finished another drink, the party was beginning to break up. Sander left the office building and walked to the garage where he had parked his car, a four-door 1965 Chevrolet. It had just started to snow. He wished the garage attendant a Merry Christmas and pulled out into the street. Traffic was very heavy at the time. Sander was six blocks from the garage when he was stopped by a policeman for reckless driving. It was quite apparent to the officer that Sander had been drinking, but rather than give him a ticket on Christmas Eve, he said that he would let Sander off if he would promise to leave his car and take a taxi. Sander agreed. The officer hailed a taxi and Sander got into it. The minute the taxi had turned a corner, however, Sander told the driver to pull over to the curb and let him out. Sander paid the driver and started back to where he had parked his own car. Upon reaching his car he proceeded to start it up and drove off.

He had driven four blocks from the street where the police officer had stopped him when he ran a red light and stuck Lowe, who was crossing the street. Sander immediately stopped the car. Lowe died a few minutes later on the way to the hospital. It was later ascertained that internal hemorrhaging was the cause of death. Sander was apprehended and charged with negligent homicide. The police medical examiner's report indicated that Sander's estimated blood alcohol concentration was between 2.5 and 3.0% at the time of the accident.

Lowe is a notorious gangster and syndicate boss who had been vying for power in the syndicate controlling the state's underworld activities. He was best known for his alleged responsibility in the Riverview massacre of five men. At the time of the incident, Lowe was carrying a loaded 32-caliber pistol which was found on his body. He had been out of jail on bond, awaiting trial on a double indictment of mail fraud and income tax evasion. Sander is a thirty-three-year-old janitor. In the building where Sander has been working as a janitor for the past two months, he was not known by many of the firm employees, but was nevertheless invited to join the party. Sander is a two-time divorcee, with three children by his first wife, who has since remarried. He was going to spend Christmas Eve with his girlfriend in her apartment. The effect of the incident on Sander was negligible; he was slightly shaken up by the impact, but suffered no major injuries. Sander has two misdemeanors on his criminal record in the past five years -- breaking and entering and a drug violation. His traffic record shows three tickets in the same space of time.

Sander was charged with negligent automobile homicide, a crime which in the state is punishable by imprisonment of one to twenty-five years.

Your job is to determine how many years of imprisonment, according to your own judgement, he should receive. After you have considered the case presented on the previous page, note below how many years of imprisonment you, as judge, would give him. Parole is beyond your jurisdiction. That is, sentence the defendant irrespective of whether or not you feel he should have opportunity for parole after a certain number of years in prison.

NUMBER OF YEARS _____

Now, would you indicate your impression of the defendant and the victim on the following scale. Circle the number which is closest to your impression of each person.

SANDERS;

extremely										extremely
	1	2	3	4	5	6	7	8	9	
favorable										unfavorable

LOWE:

extremely										extremely
	1	2	3	4	5	6	7	8	9	
favorable										unfavorable

What issues did you think were important in deciding the length of imprisonment?

What is your age? _____   Gender: Male _____   Female _____

Thank you very much for your participation. Please put this in the envelope the interviewer has so that your questionnaire will remain completely anonymous.

(4)

APPENDIX C

Laboratory Exercise 2: Gender Differences in Spatial Ability

Mental Rotations Test

Name _____

Date _____

_____

This is a test of your ability to look at a drawing of a given object and
find the same object within a set of dissimilar objects. The only difference between the original object and the chosen object will be that
they are presented at different angles. An illustration of this principle
is given below, where the same single object is given in five different
positions. Look at each of them to satisfy yourself that they are only
presented at different angles from one another.

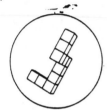 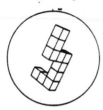 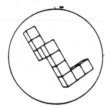 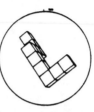 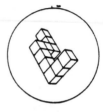

Below are two drawings of new objects. They cannot be made to match the
above five drawings. Please note that you may not turn over the objects.
Satisfy yourself that they are different from the above.

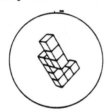 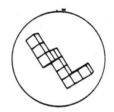

Now let's do some sample problems. For each problem there is a primary
object on the far left. You are to determine which two of four objects to
the right are the same object given on the far left. In each problem
always two of the four drawings are the same object as the one on the left.
You are to put Xs in the boxes below the correct ones, and leave the incorrect ones blank. The first sample problem is done for you.

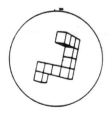 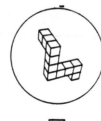 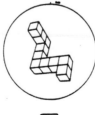  

       ☒   ☐   ☒   ☐

Go to the next page

Adapted by S.G. Vandenberg, University of Colorado, July 15, 1971
Revised instructions by H. Crawford, U. of Wyoming, September, 1979

Do the rest of the sample problems yourself. Which two drawings of the four on the right show the same object as the one on the left? There are always two and only two correct answers for each problem. Put an X under the two correct drawings.

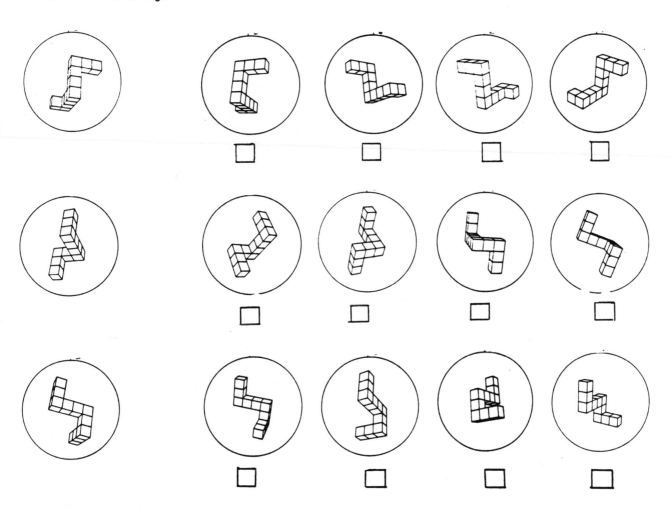

Answers: (1) first and second drawings are correct
          (2) first and third drawings are correct
          (3) second and third drawings are correct

This test has two parts. You will have <u>3 minutes</u> for each of the two parts. Each part has two pages. When you have finished Part I, STOP. Please do not go one to Part 2 until you are asked to do so. Remember: There are always two and only two correct answers for each item.

Work as quickly as you can without sacraficing accuracy. Your score on this test will reflect both the correct and incorrect responses. Therefore, it will not be to your advantage to guess unless you have some idea which choice is correct.

<u>DO NOT TURN THIS PAGE UNTIL ASKED TO DO SO</u>

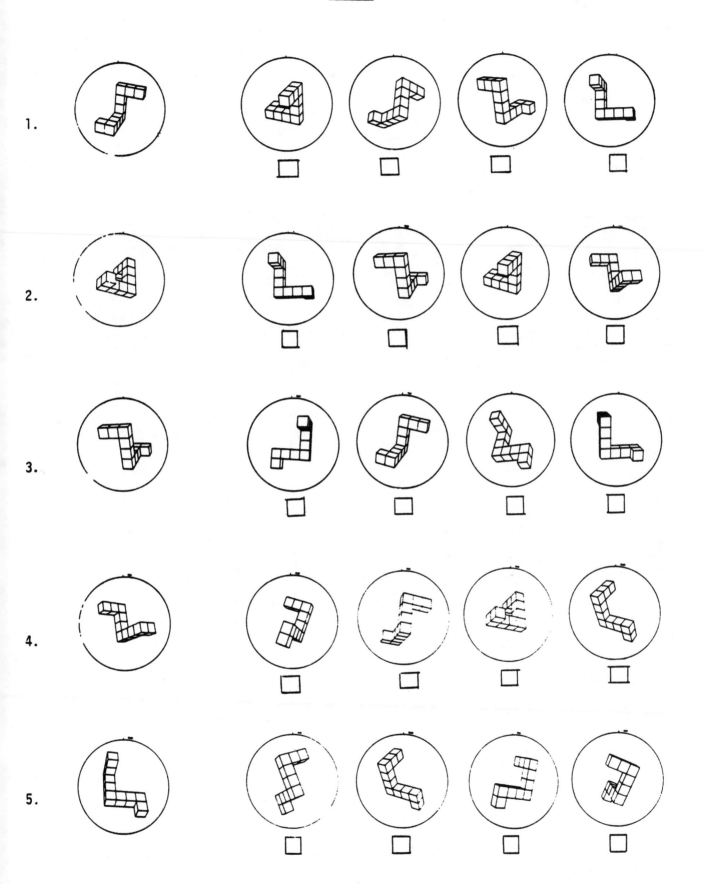

1.

2.

3.

4.

5.

GO ON TO THE NEXT PAGE

6.

7.

8.

9.

10.

DO NOT TURN THIS PAGE UNTIL ASKED TO DO SO.   STOP

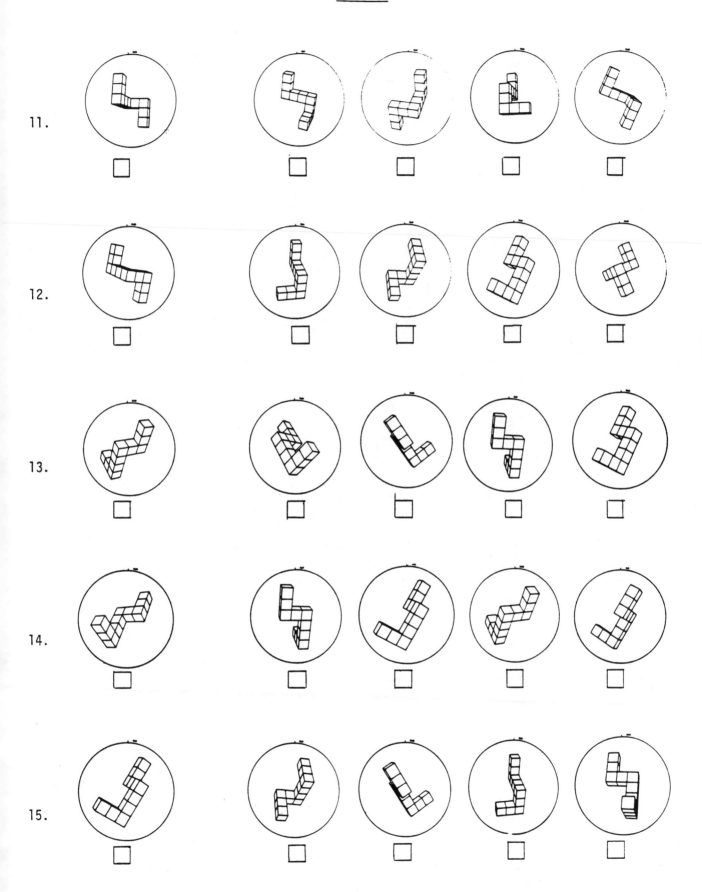

PART II

11. □ □ □ □ □

12. □ □ □ □ □

13. □ □ □ □ □

14. □ □ □ □ □

15. □ □ □ □ □

GO TO THE NEXT PAGE

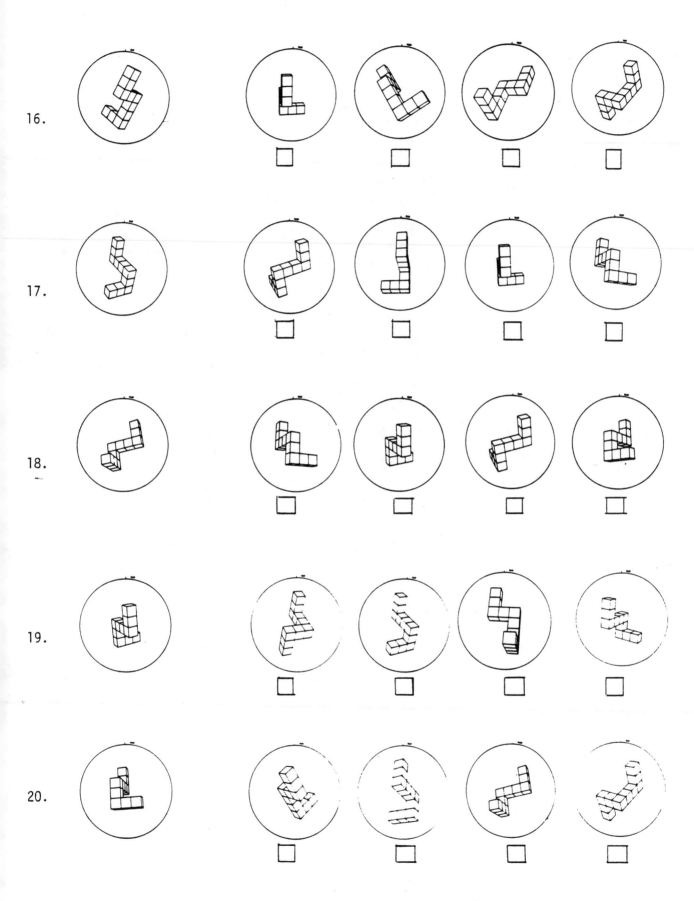

16.

17.

18.

19.

20.

DO NOT TURN THIS PAGE UNTIL ASKED TO DO SO.   <u>STOP</u>